Robert Broun

General View of the Agriculture of the West Riding of Yorkshire

Robert Broun

General View of the Agriculture of the West Riding of Yorkshire

ISBN/EAN: 9783741153259

Manufactured in Europe, USA, Canada, Australia, Japa

Cover: Foto ©Klaus-Uwe Gerhardt /pixelio.de

Manufactured and distributed by brebook publishing software (www.brebook.com)

Robert Broun

General View of the Agriculture of the West Riding of Yorkshire

GENERAL VIEW

OF THE

AGRICULTURE

OF THE

WEST RIDING

OF

YORKSHIRE;

SURVEYED

BY MESSRS. RENNIE, BROWN, & SHIRREFF, 1793;

WITH

OBSERVATIONS ON THE MEANS OF ITS IMPROVEMENT,

AND

ADDITIONAL INFORMATION SINCE RECEIVED.

DRAWN UP FOR THE CONSIDERATION OF

THE BOARD OF AGRICULTURE,

AND INTERNAL IMPROVEMENT.

BY ROBERT BROWN,

FARMER, AT MARKLE, NEAR HADDINGTON, SCOTLAND.

Oh! is there not some patriot in whose power
That best, that god-like luxury is plac'd,
Of blessing thousands, thousands yet unborn
Through late posterity? Some large of soul
To cheer dejected industry? To give
[illegible]
And teach the laboring hand the sweets of toil!
—[illegible] THOMSON.

LONDON:

PRINTED FOR RICHARD PHILLIPS, BRIDGE STREET, BLACKFRIARS;

SOLD BY FAULDER & SON, BOND STREET; REYNOLDS, OXFORD STREET; J. HARDING, ST. JAMES'S STREET; J. ASPERNE, CORNHILL; BLACK, PARRY, & KINGSBURY, LEADENHALL STREET; WILSON & SPENCE, J. TODD, YORK; SLATER & BACON, SHEFFIELD; EDWARDS & SON, HALIFAX; W. SHEARDON, DONCASTER; CONSTABLE & CO. EDINBURGH; J. ARCHER, DUBLIN; & ALL OTHER BOOKSELLERS;

BY B. M'MILLAN, BOW STREET, COVENT GARDEN.

1799.

[*Price Seven Shillings in Boards.*]

CONTENTS.

CHAPTER IV.

Mode of Cccupation.

CHAPTER V.

Implements of Husbandry.

CHAPTER VI.

CHAPTER VII.

Arable Land.

CHAPTER VIII.

Grass.

CHAPTER IX.

CHAPTER X.

CHAPTER XI.

CHAPTER XII.

Improvements.

CHAPTER XIII.

Live Stock.

CHAPTER XIV.

Rural Oeconomy.

CHAPTER XV.

Political Oeconomy as Connected with, or Affecting Agriculture.

CHAPTER XVI.

CHAPTER XVII.

CHAPTER XVIII.

APPENDIX.

CONTENTS.

INTRODUCTION.

BY MR BROWN.

THE territory which any nation poſſeſſes is the original property, fund, or capital ſtock, from whence it is ſupplied with the neceſſaries of life; to improve this capital ſtock, therefore, ought to be a primary object with every wiſe government. It is unneceſſary to deſcribe the advantages enjoyed by a country where the practice of agriculture is ſufficiently underſtood, or to mention how much that ſcience deſerves the ſupport and protection of thoſe to whom the management of public affairs is intruſted. The cultivation of the ſoil is now recogniſed as a principal ſource of national welfare, and the attention of all ranks has of late been ſo much engaged in agricultural purſuits, that a doubt can hardly be entertained, but that, when the din of war has ceaſed to prevail, the foſtering hand of the Legiſlature will be extended for its aid and encouragement.

It has excited ſurpriſe that agriculture, which, from its ſeniority, as well as utility, is entitled to a pre-eminence above commerce and manufactures, ſhould, in this country, have been hitherto left destitute of public support; while these which derived their existence from it, should, for more than a couple of centuries, have enjoyed every mark of attention and regard. In an early stage of our commerce, a public board of trade was established, and parliamentary assistance afforded, upon all occasions, to promote the infant manufactures of the kingdom. The internal improvement of the country was however undervalued and neglected, with the single exception of granting a bounty on the exportation of corn; but numberless instances might be quoted, where efficient support was withheld. In short, it seemed to be adopted as a maxim, that the hardy sons of the field were able of themselves to surmount every difficulty, but that merchants and manufacturers, like exotic plants, could not exist without legislative encouragement.

Hartlib, a respectable writer of the last century, and an eager promoter of agriculture, in the preface to his work called *The Legacy*, laments that no director of husbandry was appointed in England by *authority*. The *Museum Rusticum* likewise noticed the utility of a national establishment for regulating and superintending internal improvement, which was strongly corroborated by Lord Kaims, in his treatise called *The Gentleman Farmer;* but it was reserved to the conclusion of the eighteenth century, (an æra big with many important events), to witness the establishment of such a board; and He who was the chief cause in bringing the institution to maturity, well deserves the gratitude of every real friend to the permanent interests of this country.

It might appear presumption to attempt pointing out the benefits which will necessarily flow from the establishment of an Agricultural Board, whose measures are regulated by wise and proper principles; nor do we pretend to the possession of abilities sufficient for doing justice to such an important subject. We may be allowed, however, to remark, that their efforts will be eminently useful in procuring the removal of several obstructions to improvement, which the legal polity of England has too long sanctioned. A recommendation from them will always have weight, while the complaints of individuals are generally supposed to proceed from selfish or interested motives. The Board's opinion, of course, will also be requested in the formation of every law which affects any branch of rural œconomy.'

It will be universally acknowledged, that the first measure executed by the Board, was of the most salutary kind, and that, even if no other benefit was to be derived from the institution, a very principal object was already gained. Without ascertaining the actual state of husbandry in the ſeveral quarters of the iſland, it was impoſſible to fix upon the proper means for promoting improvement in any reſpect. This was accompliſhed by making ſurveys of all the different counties or diſtricts in the kingdom, by which means a body of *facts* was accumulated, exceeding the moſt ſanguine expectations. Theſe ſurveys, being executed by men of all profeſſions, enabled the Board to derive knowledge from a vaſt variety of ſources; and the ſcheme, happily deviſed, of circulating the original reports previous to their being re-printed in a more perfect ſtate, gave opportunity of collecting additional information, from intelligent men, both concerning the diſtrict with which they were immediately connected, and the general principles of agricultural ſcience.

Under the authority of the Board, my friends, Meſſrs Rennie, and Sheriff, and I, ſurveyed the Weſt Riding of Yorkſhire; and, during our progreſs, ſcarce a difference of opinion occurred reſpecting the matters which underwent our examination. We remained about five weeks in the diſtrict, and, during that time, uſed every means in our power to gain an intimate knowledge of the different modes in which huſbandry was carried on, as well as the general and local impediments to its improvement.

The difficulties which lay in our way in performing the bufinefs entrufted to us, may be eafily figured: Strangers to the cuftoms of the country, and not acquainted with a fingle individual in its bounds, we could not have procured the neceffary information, if it had not been owing to the liberal aid of feveral intelligent gentlemen and farmers, to whom we were recommended by Sir John Sinclair, Baronet. His letters of recommendation procured us the moft ample information; and we will always retain a grateful impreffion of the numerous inftances of attention and kindnefs fhewn us during the time we remained in the diftrict.

There is no doubt but that perfons refiding in the diftrict, might have communicated a more minute detail of feveral circumftances connected with the hufbandry thereof, than ftrangers, who, in many cafes, could only procure imperfect and contradictory accounts. Perhaps this defect was compenfated by our being apter to difcern prevailing abufes and local defects, than thofe whofe minds were familiarized with the cuftoms and ufages of the diftrict. Many things feemed to us to be of great importance, which were viewed in a different light by thofe who were refident in the country; and we certainly have faid more concerning the nature of the connection between proprietor and tenant, than a native would have thought himfelf warranted to do, or perhaps have confidered as neceffary for promoting the fuccefs of the undertaking.

When we were made acquainted with the general practice of the diftrict, in not granting leafes, it appeared ftrange to us, that perfons fo circumftanced could be expected to cultivate the ground in an advantageous way. Our attention was therefore turned to this object as deferving fpecial inveftigation. We endeavoured, in our original report, to convince the proprietors that it was impoffible they could receive the full value of their lands, under the continuance of this fyftem, and pointed out the many happy confequences which would accompany the granting free and open leafes. We are forry to learn, our arguments on this head have given offence to a great number of that body, which was a circumftance very foreign to our intention; but, convinced of their rectitude, we have, in this re-printed copy, rather enlarged than contracted our original remarks. To us, it would feem as incongruous to tye a man's legs together, and then order him to run, as to fuppofe, that improvements are to be made by a farmer, without the fecurity of a leafe. The great charm which fets induftry every where in motion, is the acquifition of property, and the fecurity of it when acquired. Where tenants hold by a precarious tenure, and are removable at the will of the proprietor, or after a fhort period, then undoubtedly their labour will be ſpiritleſs and languid, as they have no inducement to enter upon improvements, when they have no certainty of enjoying the immediate benefit.

It is now proper to ſay a few words concerning this ſecond edition of the ſurvey.

When the Board ſignified their deſire, that we ſhould undertake the taſk of preparing the work for re-publication, application was immediately made to almoſt every perſon, who had formerly favoured us with intelligence, and they were particularly requeſted to point out any errors in the original copy reſpecting facts, which we conſidered as of the utmoſt importance. In conſequence of theſe applications, a good deal of additional information was received, which is incorporated with the text, where it did not militate againſt the ſentiments formed in our progreſs. The copies, returned to the Board with marginal remarks, were alſo conſulted; and every thing favourable or unfavourable to our opinions has been inſerted, either in the body of the Work, in the Appendix, or by way of Notes. In ſome caſes the latter were ſo hoſtile, that we have thought it neceſſary to follow them with ſuitable anſwers.

The arrangement, ſuggeſted by the Board, has been uniformly adhered to, unleſs in ſome few articles of leſſer conſequence, which we judged inexpedient to diſcuſs.

We are aware, the manner in which we have treated the different ſubjects, is rather contrary to the rules laid down by the Secretary of the Board, in his introduction to the Suffolk ſurvey; but, with all due reſpect to the ſuperior talents of that gentleman, we muſt conſider what he ſays as not applicable to the buſineſs. If his rules were ſtrictly adhered to, a ſurvey would be no more than a collection of ſtatiſtics; nay it would not contain the whole ſtatiſtics of a county or diſtrict, for if another county poſſeſſed the ſame particulars, then it was improper to inſert them in what he calls a local ſurvey. We are clear that nothing ſhould be treated in the ſurvey of any diſtrict, but what is connected with the huſbandry thereof; but certainly if the fact needs to be illuſtrated by arguments, they are not out of place, merely becauſe the ſame arguments might be uſed reſpecting the huſbandry of another diſtrict. The perfection of hiſtory is to develope the cauſes which have produced the events recorded, and to accompany the narrative with ſuitable obſervations; but if Mr Young's rules were applied to a hiſtorical performance, every article relative to the ſtate of other nations, ought to be expunged as being out of place, and the work would degenerate into a mere body of dry annals, without furniſhing inſtruction or amuſement.

But let us ſee what ſort of a work the Suffolk Survey would have been, had carrots, cabbages, polled cows, and poors houſes been common in the conterminous counties. If Mr Young had adhered to his own rules, he behoved juſt to have mentioned thoſe articles without enlarging upon them, becauſe the chapter or ſection might be equally applicable to the huſbandry of other counties. Shall the chapter upon leaſes, for inſtance, be juſt entered upon and left off immediately, becauſe the want of them is a grievance, which affects a great part of the kingdom, or ſhall a general ſubject be neglected merely becauſe the whole iſland is intereſted in its diſcuſſion. Such a conduct would be as prepoſterous, as that of a phyſician would be, who refuſed to preſcribe for a patient, becauſe the *recipe* might be equally applicable to the caſe of a perſon in the next village afflicted with a ſimilar diſorder.

Though the leading part of a ſurvey is to repreſent the actual ſtate of huſbandry in the diſtrict, it may be queſtioned whether the public will derive ſo much benefit from this branch of theſe performances, as from a faithful deſcription of the obſtacles to improvement, and the means by which they can moſt judiciouſly be removed. It is in theſe departments the ſurveyors have the fitteſt opportunity of benefiting the public, or of communicating to the Board uſeful information. If the huſbandry of the kingdom was uniformly good, we acknowledge there would be little occaſion for ſaying much reſpecting thoſe matters; but in the preſent ſtate of rural affairs, we, with ſubmiſſion, contend, they ought principally to engage the attention of the ſurveyors.

It is from a compariſon of the ſentiments of the different ſurveyors, upon ſimilar ſubjects, that the Board can be enabled to form a true idea of the preſent ſtate of Huſbandry in Britain, or be guided in their deliberations upon the means for promoting internal improvement. Freedom of enquiry ought to be encouraged, as the only way of arriving at truth; for if the ſurveyors are tied down by arbitrary rules, the opinion of one man may as well be conſidered as infallible, or taken as a criterion for aſcertaining the ſtock of knowledge in the kingdom. We mention thoſe things, becauſe our ſurvey is drawn up on quite different principles, from theſe pronounced by Mr Young as neceſſary to conſtitute a county report.

It is certainly neceſſary to apologize for the many errors which prevail in this work. Diſtance from the preſs and a crowd of other avocations prevented that correctneſs of compoſition, which is to be found in ſeveral works of the like nature. But perfection in compoſition is not to be expecied from thoſe engaged in the *practice* of rural ſcience, nor will the want of it be laid to their charge as a crime. According to the Reverend Mr Harte 'the plain practical author pays his 'little contingent to the republic of knowledge 'with a bit of unſtamped real bullion, whilſt the 'vain glorious man of ſcience throws down an 'heap of glittering counters, which are gold to the 'eye, but lead to the touch-ſtone.'

We truſt that our obſervations will be candidly conſidered, and that unintentional defects will be forgiven. We are not conſcious of having miſrepreſented a ſingle fact, or of having offered an opinion, which, to the beſt of our judgement, would prove diſadvantageous to the public. Others might have executed the work with greater ability, but we muſt be pardoned for declaring that few could have been more anxious to preſent to the Board a report, which would communicate a faithful account of the preſent ſtate of Huſbandry in the diſtrict, and at the ſame time deſcribe the obſtacles to improvement, and how they might be removed.

PRELIMINARY OBSERVATIONS.

A Confiderable number of remarks being returned to the Board, upon the firft edition of this report, it is judged neceffary to prefent the greater part of them in this amended copy of the work, in order that the Public, from a view of both fides of the queftion, may be enabled to judge for themfelves. None have been fuppreffed, however hoftile to the fentiments of the furveyors, which were of the fmalleft importance, except thofe upon the article of tithes, which are left out for reafons to be afterwards mentioned: Indeed we are more apprehenfive of being cenfured for admitting a number of obfervations, apparently dictated by a petulent capricioufnefs, than for making a partial felection of the marginal information tranfmitted to us.

We have thought it moft regular to infert the remarks at the conclufion of the feveral chapters or fections with which they are connected, and the utmoft care has been beftowed to diftinguifh the different places to which they refer. After all, from the great quantity of new matter received fince the printing of the original report, we will not warrant that they are always exactly marked.

If any error has happened in copying the proper names, eſpecially thoſe contained in the Appendix, we truſt that it will be excuſed by thoſe who liberally favoured us with ſuch a variety of local information.

AGRICULTURAL SURVEY

OF THE

WEST RIDING OF YORKSHIRE.

GENERAL DESCRIPTION OF THE WEST RIDING OF YORKSHIRE.

YORKSHIRE is by far the largest county in the kingdom, and is divided into three Ridings, viz. the East, West, and North; each of which is as extensive as the generality of other counties. Mr Thoresby, in his History of Leeds, says, that in the division of England by the Saxons, for the better government of it, there were these parts, viz. Tythings, Hundreds or Wapentakes, and Trithings, or Ridings, which thus differ: Tythings consisted of ten families, subjected to the care of the Overseer or Tything-man, who was to be answerable for the behaviour of the masters of those families, as they were of their children and servants. Ten of those Tythings made an Hundred or Wapentake, which last was so called because the governor of it was put into his place, and held up a weapon, i. e. a spear, and the elders of the Tythings admitted him, by tacking or touching their spears with his, as a token of their subjection to him. Ridings or Trithings were the third part of a county, be it greater or leſſer, and to them were appeals made in caſes not determinable in the Wapentakes.

This county, in the time of the Britons, was inhabited by the *Brigantes*, whoſe territories included the preſent counties of Cumberland, Durham, Lancaſter, Weſtmoreland, and York. During the Saxon government, it made part of the kingdom of Northumberland, till the Weſt Saxon kings ſubdued the other ſix kingdoms, and formed the whole of England into one monarchy.

It is only one of the diviſions of Yorkſhire which we are now to deſcribe, viz. the Weſt Riding; and, without all diſpute, it is the moſt important of the three. It contains not only a large quantity of valuable ground, well adapted to the different purpoſes of huſbandry, but alſo, in its bounds, are carried on large and extenſive manufactures. In a word, whether it is conſidered with reſpect to magnitude, fertility of ſoil, local advantages, manufactures, or population, it will be found deſerving the moſt minute attention, and worthy to be ranked with any province in the kingdom.

CHAPTER I.

GEOGRAPHICAL STATE AND CIRCUMSTANCES.

SECT. 1.—*Situation and Extent.*

THE West Riding of Yorkshire is situated nearly in the centre of the kingdom; and although an inland district, yet, from numerous rivers and canals, possesses all the advantages of a maritime province. It is bounded on the East by the Ainsley of York, and the river Ouse, which river separates it from the East Riding; on the south, by the counties of Nottingham and Derby; on the west, by the counties of Westmoreland, Lancaster and Chester; and, on the north, by the North Riding; and is 95 miles in length from East to West, 48 miles in breadth from South to North, and about 320 in circumference, containing 2450 square miles, or 1,568,000 statute acres.

SECT. 2.—*Divisions.*

THE West Riding is divided into nine Wapentakes, viz. Aghridge, Barkston, Claro, Morley, Osgoodcross, Skirack, Staincliffe, Strasford, and Staincross. It contains 175 parishes, several of which are of great extent; 28 market towns, the chief of which are Leeds, Sheffield, Wakefield, Halifax, Bradford, Huddersfield, Barnsley, Selby, Skipton, Settle, Snaith, Ripon, Pontefract,

Knarefborough, Rotherham, and Doncafter, befides a great number of populous villages.

The general Eafter feffions for the whole Riding are held at Pontefract, and continue for a week. The midfummer feffions are opened at Skipton, and when the bufinefs in that quarter is gone through, the Magiftrates adjourn to Bradford. The Michaelmas feffions are firft held at Knarefborough, then adjourned to Leeds, and afterwards to Doncafter. The Chriftmas feffions are held at Wetherby, Wakefield, and Rotherham. Pontefract may therefore be confidered as the county town, though the records of the feffions, and regifters of the landed property are kept at Wakefield.

SECT. 3.—*Climate.*

As the Riding is of great extent, and contains large tracts both of mountainous and low land, the climate, of courfe, varies much. Upon the whole, however, it is moderate and healthy, except near the banks of the Oufe, where, from lownefs of fituation, damps and fogs fometimes prevail. The harveft over the greateft part of the diftrict is comparatively early, commencing ufually before the middle of Auguft, and, backward feafons excepted, is finifhed by the end of September; but, in the weftern parts, it is at leaft a fortnight later than about Pontefract and Doncafter. The average gauge of rain, at Sheffield, is 33 inches in a year, which is about a medium betwixt what falls in Lancafhire, and on the eaftern coaft.

SECT. 4.—*Soil and Surface.*

THE face of the country is ſtrongly irregular. In the weſtern and northern diviſions a conſiderable portion is hilly and mountainous; though in theſe ſituations it is interſected with numerous vales, carrying graſs of the richeſt quality; but the middle and eaſtern parts are generally level, having no more eminences than what ſerve to variegate the proſpect.

The whole arable land is nearly incloſed with ſtone walls and hedges, which are kept in good condition; and there are few open fields, but where the ground is common or waſte.

The nature and quality of the ſoil, in this extenſive diſtrict, differs materially. There are all ſorts, from the deep ſtrong clay and rich fertile loam, to the meaneſt peat earth; and probably it contains all the different varieties that are to be found in the iſland. Vicinity to great towns, and ſuperior culture have, no doubt, rendered a conſiderable part fertile and productive that was originally barren; but a large proportion of the diſtrict is of a quality naturally favourable to the purpoſes of good huſbandry, and, under a proper ſyſtem of management, will amply repay the farmer for whatever trouble and expence he beſtows on its cultivation.

SECT. 5.—*Minerals.*

THERE are numerous mines of coal, lime, ironſtone, and lead, and ſome copper, in this diſtrict, which have been wrought for ages paſt, and may, in ſome places, be ſaid to be inexhauſtible. At Graſſington the lead mines are numerous and valuable, but they are now wrought with leſs advantage than formerly, owing to the want of a freſh level, which can only be done by the Duke of Devonſhire who is Lord of the manor. We believe his Grace formerly took one ſeventh for his dues, but of late, in freſh bargains, he demands one fifth, which is far too high. If he was to reduce his claim to one ſeventh again, he would be a conſiderable gainer.

SECT. 6.—*Water.*

THE Weſt Riding is remarkable for the number of its great and navigable rivers: *1ſt*, The Ouſe which takes this name a few miles above York, being formerly called the Eure, and in its courſe to the Humber receives all the other rivers that run through the diſtrict. *2dly*, The Don, which is navigable nearly to Sheffield, and of great advantage to the trade of that neighbourhood. Over this river, betwixt Snaith and Thorn, there is a wooden bridge which turns upon a pivot, and affords a paſſage for the numerous ſhipping employed in the inland trade. *3dly*, The Calder, which flows along the borders between this Riding and Lancaſhire, and running in an eaſtern direction falls into the Aire, five miles below Wakefield. *4thly*, The Aire a large river iſſuing from the mountain Penigent; which, with the aid of canals, is navigable to Leeds, Bradford, and Skipton. *5thly*, The Wharfe which has its riſe at the foot of the Craven hills, and after a courſe of more than 50 miles acroſs the Riding, keeping for a great way an equal diſtance of 10 miles from the Aire, diſcharges itſelf into the Ouſe. Beſides theſe principal rivers there are many of leſſer importance.

CHAPTER II.

STATE OF PROPERTY, AND THE TENURES UPON WHICH IT IS HELD.

TO ascertain the ſtate of property in this diſtrict, and to deſcribe the different tenures, upon which it is held, would have required conſiderably more time than we could have devoted to theſe objects. Theſe are parts of an agricultural ſurvey which it is impoſſible for ſtrangers to diſcuſs with ſuch accuracy and preciſion, as could have been done by perſons more intimately acquainted with the uſages, cuſtoms, and practices of the diſtrict ſurveyed. Perhaps, after all, the two points which occupy this chapter are of as little importance as any other head of this work, and their inveſtigation, however much it might gratify curioſity, can be of little or rather of no material utility.

A conſiderable part of the Weſt Riding is poſſeſſed by ſmall proprietors, and this reſpectable claſs of men, who generally farm their own lands, are as numerous in this diſtrict as in any other part of the kingdom. They are uſeful members of the ſtate; they are attentive in the management and cultivation of their lands; and they form an important link in the chain of political ſociety. There are likewiſe a great number of extenſive proprietors, ſuch as the Duke of Norfolk, Earl Fitzwilliam, &c. whoſe annual income it is unneceſſary, and at the ſame time it would be improper, to ſtate. Few of the large proprietors reſide upon their eſtates, at leaſt for a conſiderable part of the year, and the management of them is moſtly devolved on their ſtewards, who, from being early trained to buſineſs, are generally intelligent, active, and induſtrious men.

The greateſt part of the Riding is freehold property, which is evident from the aſtoniſhing number of freeholders reſiding in it, the number of copy-holders, or thoſe who hold by a copy of court-roll, is alſo conſiderable. A good deal of land likewiſe belongs to the Archbiſhop, Colleges, Deans, Prebends, and other church dignitaries; and the inferior clergy, in conſequence of incloſure bills, are accumulating landed property every year.

CHAPTER III.

BUILDINGS.

SECT. 1.—*Houses of Proprietors.*

TO describe the houses of the proprietors is perhaps foreign to the business of the agriculturist. Suffice it to say, that the West Riding contains a number of magnificent and elegant houses belonging to the nobility and gentry who have property in it. Without pretending to enumerate them, we shall content ourselves with saying, that Wentworth House the property of Earl Fitzwilliam, is without any doubt one of the largest and most magnificent in the kingdom.

SECT. 2.—*Farm Houses and Offices.*

THE farm houses and offices are, in most cases, very inconveniently situated, being generally crowded into villages or townships, and not placed on the lands the farmer has to cultivate. Whatever necessity for this practice arose from the circumstances of former times, when property was insecure, and exposed to ruinous depredations, it is obvious there can be none for it now, when these circumstances are wholly removed. It is equally clear, the nearer the houses of the farmer are to the lands he occupies, the more work may be performed, and consequently his operations will be carried on not only with greater convenience, but also at less expence. These things, we are happy to ſay, are now attended to more than formerly, although much room is ſtill left for further improvement.

Here we beg leave to notice the ſuite of farm-offices, lately erected by the right honourable Lord Hawke, which affords an elegant pattern for his neighbours. His Lordſhip has built, for his own uſe, a large farm yard, conveniently formed and ſituated, with a threſhing machine, a mill for grinding rapecake, ſtables for 25 horſes and 32 oxen, beſides cow ſheds, barns for hay, corn, &c. The whole is ſurrounded by walls nine feet high, and divided by the barns, ſtables &c. into four yards, two of which have ponds, beſides the pumps. The ſtables for the horſes are placed on the Eaſt and Weſt ſide of the farm yard which is free from buildings on the South, and ſheltered on the North by the barn and ox houſes, which ſeparate it from the principle ſtack yard. This yard is divided from the two others by open hay barns, tiled with ſlate eaves and with chimnies alſo of brick to let out the ſteam. The average of the boarded granaries amounts in length to an hundred and ſixty feet, and in breadth to 21 feet. There are trap doors contrived in them to let down the corn, when ſacked, into waggons which may be loaded and locked up at the ſame time. The corn in the yard is ſtacked on wooden frames placed on ſtone pillars and capes. When we ſaw it, Lord Hawke propoſed to make further improvements on it, and to build a houſe for his ſteward. The whole indeed, forms a complete elegant and convenient ſuit of farm-offices, covering from one to two acres of ground, and is in every reſpect becoming a nobleman who juſtly conſiders the cultivation of the earth as the moſt uſeful and neceſſary of human employments.

As nothing contributes more to promote, the happineſs and comfort of a farmer, than to have his farm ſteading or offices properly conſtructed and conveniently ſituated, we ſhall here ſtate our opinion on the manner in which theſe buildings ſhould be placed, when they are intended for the uſe and accommodation of the practical farmer.

The farm-houſe and offices ſhould be placed as near as poſſible in the centre of the farm, provided good water can be got in plenty, which ought always firſt to be enquired after. The farm yard or fold yard ſhould be a long ſquare proportioned to the ſize of the farm, and the number of buildings intended to be erected. The barns ought to be placed on the weſt ſide of the yard, the ſtables and byres for horſes and milk cows on the ſouth, byres for feeding turnip cattle, and houſes for lodging huſbandry utenſils on the eaſt, and on the north open ſhades, where cattle that are wintered in the ſtraw yard, may ſhelter themſelves during bad weather.

This affords complete conveniencies of all kinds, and keeps every thing within the reach and ſight of the farmer, which is an object of great importance.

The dwelling houſe for the farmer, we think, ſhould be placed at a ſmall diſtance, ſay 20 or 30 yards from the farm yard, which both removes his family from the filth and naſtineſs which muſt neceſſarily prevail where cattle are kept, and contributes to prevent accidents from fire.

Where the farmer employs a machine for threſhing out his corn, we would recommend that the barn in which it is placed, ſhould be extended into the ſtack yard, which renders the houſing of the ſtraw much more convenient than if the machine was placed in the ſtreight line of the farm yard; a row of cottages for farm ſervants, ſhould be built at a little diſtance, ſay a hundred yards, from the ſuite of offices.

We had occaſion to notice the great ſize of many barns

presently used in the West Riding, which, in our humble opinion, are attended with an unnecessary expence. The building such edifices at first is not only a great burthen upon the farmer, but the interest of the money originally laid out, and the sums required for keeping them in repair must be great, while at the same time these unnecessary expences are productive of no real benefit to the farmer. The reason assigned to us for having such large barns was, that as much of the crop might be housed as possible, when taken from the field. We can perceive no utility from this practice, as corn can never be kept so well in a house as when properly stacked in the yard. It will always be found drier and healthier in that situation than when kept long in the house, which it must necessarily be wherever large barns are used; besides, in backward seasons corn can be got much sooner ready for the stack than the barn, and it is an important article of farm œconomy to have it as soon out of danger as possible.

It is said housing of corn saves expence. This we doubt, as it will take as many people to put it into the barn in harvest, as afterwards, and the difference of expence betwixt harvest and common wages will build it in the yard; at any rate, the expence of the barns, and the danger of the corn turning mouldy in them, far more than exceed every advantage that can be derived from this practice.

We also noticed, that when corn was built in the yard, the stacks were of an oblong form, whereas we think it cannot be built in a more easy and convenient manner than in round ones. These may be made of any size the extent of the farm requires, and from their shape and construction the air penetrates with greater facility into the heart of the stack than when built in the oblong form.

Perhaps a good deal of unneceſſary trouble is beſtowed upon covering both hay and corn ſtacks, as the ſtraw is laid on in great quantities and with as much accuracy as if it were thatched for a dwelling houſe; while the roping is as ſtrongly applied as if the ſtacks were to ſtand for twenty years. We admit that corn ought always to be properly ſecured, and are far from condemning theſe practices becauſe they are accurate, but we think the preſent mode of covering ſtacks an unneceſſary waſte of labour and expence, and that the corn will be as well defended from the weather if half the trouble was ſaved.

SECT. 3.—*Cottages.*

THERE is a great want of dwelling houſes for huſbandmen and labourers; and this deficiency may be traced to the poor laws for its ſource. The farmer, from a dread of heavier rates falling upon him, keeps as few houſes as poſſible; and hence, almoſt the whole of the farm ſervants are young unmarried men, who have board in the houſe; while thoſe thoſe that are ſtyled day-labourers, reſide in the villages. This practice is very troubleſome to the farmer: it decreaſes the number of people employed in huſbandry; and has, for its certain attendant, a great riſe of wages.

We venture to recommend, that proper houſes ſhould be built for farm ſervants, contiguous to every homeſtead. This will not only promote the welfare and happineſs of that claſs of men, by giving them an opportunity of ſettling in life, which is not at preſent an eaſy matter, but will alſo be highly beneficial to the farmer himſelf, as he will at all times have people within his own bounds, for carrying on his labour; and have them of that deſcription, that are generally eſteemed moſt regular and careful. (*a*)

We alſo recommend that married farm ſervants ſhould receive their wages, or at leaſt the greateſt part of them, in the produce of the ſoil, which would be advantageous to that claſs of people, and not detrimental to their maſters. Under this mode of payment, they are always certain of being ſupplied with the neceſſaries of life, and a riſe of markets does not affect them; whereas, when the wages are paid in money, they are expoſed to many temptations of ſpending it, which their circumſtances can but ill afford, and during a riſe of prices are often reduced to the greateſt ſtraits. In Scotland, farm ſervants are uſually paid in this manner; they receive certain quantities of oats, barley, and peaſe, have a cow ſupported during the whole year, and a piece of ground for raiſing potatoes and flax. We are aware how difficult it always is to introduce new cuſtoms, but we are ſo ſenſible of the beneficial conſequences accompanying this mode of paying farm ſervants, that we earneſtly wiſh it was adopted over the whole kingdom.

NOTES on CHAPTER III.

Section 3.

(*a*) The building of cottages contiguous to the farm offices, would be a great convenience to the farmer, and of greater advantage to the community. *T. H.*

Cottages with 3 or 4 acres of land, are very much wanted. From the want of a little land laid out to cottagers in every parish, there is a most *crying* scarcity of that almost indispensible necessary for the rearing of children, MILK. Even in the most plentiful and fertile parts of the country, farmers think it their interest to give their spare milk to the pigs, and they too generally discourage the letting of bits of grass land to cottagers; whether for fear of rendering them more independent of themselves, or that landlords should discover that cottagers can give higher rents, or from what real cause I know not; however, the beneficial effects of this plan to land owners, and the poor in the few parishes, as instances where it fortunately obtains, are so great and manifest, that it is matter of astonishment to me, it has not been more generally adopted. A number of useful milk cows, kept amongst the poor labourers, has a tendency to diffuse the blessings of plenty, property, and a love of order, in a manner most beneficial to the community; and it is a kind of trade, (that of milk) which a poor man and his wife know best how to manage among their poor neighbours, so that a very few cows in their hands would supply a pretty large village.

W. P.

CHAPTER IV.

MODE of OCCUPATION.

SECT. 1.—*Size of Farms.*

THE majority of farms are comparatively ſmall, and few are of that ſize as would be conſidered in other parts of the kingdom as large ones. Upon the arable lands we heard of none exceeding 400 ſtatute acres, and for one of that extent there are a dozen not fifty acres. In the graſs diviſion of the county they are ſtill ſmaller, and we often heard the occupier of a hundred acres of ground ſtyled a great farmer.

Various cauſes might be aſſigned for land in the Weſt Riding being occupied in ſuch ſmall portions. Manufactures being carried on to ſuch extent has naturally occaſioned capitals to be laid out in trade, which, in other counties, would be employed in agriculture; and wherever this is the caſe the occupiers of the ground will generally be found deſtitute of ſtock for cultivating the ground in an advantageous manner, and defective of knowledge in the ſcience they practice. We hazard this as a general obſervation, without applying it to the farmers of the Weſt Riding, many of whom are as enlightened and liberal as any of their profeſſion in the iſland.

The proper ſize of a farm, is a queſtion upon which theoriſts have often diſputed. In our inquiries, we wiſh to be regulated by practical principles; and although we are fully convinced, that a farm of a proper extent, ſuited to the capital and abilities of the poſſeſſor, operates as a ſpur to activity and diligence, yet we are not advocates for a ſyſtem that would monopolize the lands of any country, by throwing them into the hands of a few.

An improved ſyſtem of huſbandry, requires that the farm upon which it is to be carried on ſhould be of ſome extent, or elſe room is not afforded for the different crops neceſſary to complete a perfect rotation of management. The farmer, who practiſes huſbandry upon proper principles, ſhould not only have his fields under all ſorts of grain, but likewiſe a ſufficient quantity of graſs and winter crops, for carrying on his ſtock of cattle and ſheep through all the different ſeaſons of the year. By laying out land in this ſtyle, the economy of a farm is ſo regulated, that while improvements progreſſively go forward, too much work does not occur at one time, nor occaſion for idleneſs at another. This, when the expences of farm-culture are ſo extravagant as at preſent, deſerves particular attention; but cannot, in the nature of things, be juſtly and accurately arranged, where the farm is of ſmall ſize.

It may be imagined, that the arrangement of farm-labour, and the cultivation of the ground, whatever the ſize of the farm may be, is but a rule-of-three queſtion; and that the ſmallneſs of the poſſeſſion only reduces the ſcale upon which improvements are to be carried on. This may in part be true; but will the reſult of the queſtion be favourable to improvements? Upon 50 acres, labour may not be afforded for half a team; the incloſures would perhaps be a few acres, and the farmer would go to market and buy a ſingle beaſt, thereby affording opportunity for ſpending half the year in idleneſs, waſting the ground by a number of fences, and occaſioning more expence than the whole profit would repay. These things are the neceſſary conſequences of arranging farm management like an arithmetical queſtion, and are great drawbacks upon the profits of farming.

Beſides, an improved ſyſtem of huſbandry requires the farmer ſhould be poſſeſſed of an adequate ſtock, a thing in which ſmall farmers are generally deficient. It is an old proverb, the truth of which we have too often ſeen exemplified, "that the poor farmer is always a bad one." Allowing he has knowledge, he cannot reduce it to practice, for want of the neceſſary means. The ſmallneſs of the Weſt Riding farms, and the precarious ſituation of the farmer's condition, ariſing from want of leaſes, as well as the trammels under which he is obliged to work, have, in a great meaſure, thrown capitals into another line. Unleſs theſe circumſtances are altered, perſons of abilities, and poſſeſſed of ſtock, will be induced to deſpiſe the profeſſion, and agriculture will not be carried on in its moſt improved ſtate.

With regard to the queſtion, whether large or ſmall farms are generally beſt managed? we apprehend very few words will ſuffice. Who keeps good horſes, and feeds them well? Who makes the completeſt fallow, takes the deepeſt furrow, and ploughs beſt? Who has the greateſt number of hands, and ſufficient ſtrength for catching the proper ſeaſon, by which the crop upon the beſt of grounds is often regulated? Who purchaſes the moſt manure, and raiſes the weightieſt crops? We believe, in the general, theſe queſtions muſt be anſwered in favour of the large farmer. If ſo, it follows that the prevalence of ſmall farms in the Weſt Riding of Yorkſhire retards its improvement.

It is a popular doctrine, that large farms are unfriendly to population, and that they ought to be diſcouraged. We ſuſpect this doctrine is founded in prejudice, and will not ſtand the teſt, if accurately examined. No doubt, if farms are increaſed in ſize, the number of farmers is leſſened; this is granted: but with regard to the great ſcale of population, we are clearly of opinion it is not affected. If a more ſuperior practice is carried on upon a large farm than a ſmall one, this muſt be accompliſhed by employing a greater number of hands. What, therefore, is loſt in one claſs, is gained in another. Beſides, we have often noticed, that upon large farms moſt married ſervants are kept, which affords encouragement to the increaſe of population. Upon a ſmall farm, from 50 to 100 acres, what is the farmer to do? he has not ſufficient buſineſs for employing his attention, and the ſmallneſs of his poſſeſſion will not allow him to be idle. He therefore muſt work with his hands, which brings the queſtion preciſely to the ſame iſſue, as if all work was performed by hired ſervants; independent of the arguments we have adduced, that more work is executed, and more hands employed, upon a large farm, than upon the ſame extent of land divided into ſmall ones.

It has given us ſurpriſe to obſerve many perſons taking it for granted, that by increaſing the ſize of a farm you neceſſarily decreaſe the number of the people, without conſidering that if the management is equal in every reſpect, the population muſt be exactly the ſame, with the exception of one or two farmer's families. They tell you that cottages are pulled down, whereas the large farmer has occaſion for more cottages than the ſmall farmer, as he cannot keep ſo many houſe ſervants, and is often under the neceſſity of building new houſes, in order that the number of ſervants he keeps may be accommodated. An attentive obſerver will ſmile at the doleful pictures often exhibited by ſuch alarmiſts, which, to do them juſtice, are not original ones, as they have been borrowed from former times. In a word, wherever work is carried on, it muſt be done by employing hands, and where the work is executed in the moſt perfect manner, the greateſt number of hands muſt be employed. If the ſyſtem carried on upon the premiſes is improved, the population muſt of courſe be increaſed; the one is the cauſe, the other is the effect, and practice and daily experience juſtifies the concluſions we have drawn.

SECT. 2.—*Rent.*

IT is difficult for us to ſay what may be the real rent of land. We could not, with propriety, puſh the farmer upon this point, when he was ignorant what uſe we were to make of his anſwer; and even where we got ſufficient information of what was paid the landlord, we found there was a long train of public burthens, over and above, which could not be eaſily aſcertained. There is, in the firſt place, the land tax, which is uniformly paid by the tenant, and generally amounts to 1s. per pound upon the real rent. 2dly, The tithes, which are levied in ſo many various ways, that it is impoſſible to ſay what proportion they bear to the pound rent, much depending upon the actual ſtate of the farm, and not a little upon the character and diſpoſition of the drawer. Upon arable lands, where they are annually valued, the payment of money may be from 5s. to 8s. per acre, in ſome caſes more. 3dly, The roads, the expence of which to the tenant is about L. 7 per cent. upon the rent. 4thly, The poor rates, for which no fixed ſum can be ſet down. The loweſt we heard of was 18d. in the pound; and the higheſt 8s.; but from the very nature of the tax they are continually fluctuating, and ſince our ſurvey was made are greatly increaſed. 5thly, The church and conſtables dues, which are about 1s. in the pound.——From all theſe things it may be ſuppoſed, that in many places the ſums payable by the farmer the church, the public, and the poor, are nearly as great as the nominal rent paid to the landlord. It will appear ſurpriſing to many, that rents are higher for graſs fields than for thoſe under the plough (*a*). This is however actually the caſe, and we account for it in the following manner. When in graſs, few or no tithes are paid, at leaſt the burthen is comparatively light. The want of leaſes, the reſtrictions commonly impoſed, and the payment of tithes do not operate half ſo ſeverely upon the grazier as upon the corn farmer. The graſs farmer has few improvements to make; he goes on in the ſame courſe from year to year; and the want of a leaſe, though it keeps him from the certainty of poſſeſſion does not hurt him ſo far as to cramp his operations (*b*). At Settle and Skipton, we found that land let ſo high as 40s. and 50s. per acre, while, from the beſt accounts we could receive in the corn country, 20s. and 30s. was then conſidered as a high rent, and in many places it was much lower (*c*).

NOTES on Sect. 2.

(*a*) This is true in the caſe of land of the beſt quality. Inferior land is uſually improved by being brought into a good courſe of tillage. *T. York, Eſq.*

(*b*) The rent of paſture and meadow land is higher, I believe, in moſt countries, than that of arable, and for reaſons ſimilar to thoſe here given. *Anonymous.*

But if the peculiar burdens affecting corn land were removed, this would not be the caſe. *R. B.*

(*c*) Graſs products have of late been at a higher proportional price than corn. Foreigners can frequently underſell us in our own corn markets; not ſo in thoſe for graſs products: Fat beef mutton, butter, milk, &c. are bad articles for importation. The tithe will ever be an inducement to turn the balance from corn to graſs in many caſes. *A Yorkſhire Freeholder.*

SECT. 3.—*Tithes.*

THIS is an important article, which well deferves the minuteft confideration of the Board of Agriculture. For reafons to be afterwards mentioned, we decline inveftigating the confequences attending the payment of tithes, whether they are confidered as a part of the tenant's rent, operating in direct proportion to his induftry or abilities, or as a tax originally impofed for certain purpofes, which circumftances have now totally changed. That it may be feen that the fuppreffion of what we formerly faid againft the payment of tithes, either by an annual valuation, or by an exaction in kind, does not proceed from any change of principle, or alteration of fentiments, we fubjoin an extract of a letter from Sir John Sinclair refpecting this part of our furvey, which we are authorifed to publifh in our own vindication.

" In drawing up this work, there is only one reftric-" tion, which I wifh to impofe upon you; it relates to " the payment of tithes, a fubject of great delicacy and " importance, which regards only the fifter kingdom, " confequently it is a point with which we North Britons " have no particular occafion to interfere. I wifh, " therefore, that in your report, any particular difcuffion " of that fubject may be avoided."

After the reftriction thus laid upon us refpecting this article, it would be improper to fay more than that the real intereft of the country is concerned in having tithes regulated as foon as poffible.

In a moral point of view, every well difpofed perfon muft lament that the collection of a tax, originally defigned for the fupport of religion, fhould now be the means of creating difrefpect for its minifters. There are no arguments neceffary to prove, that where the clergyman differs with his parishioners upon this subject, the usefulness of his office is totally frustrated; which makes not only the practice, but even the profession of religion be disregarded.

Sect. 4.—*Poor's Rates.*

The expence of supporting the poor is another burden on the possessors of land, which has of late greatly increased. In a district, such as the West Riding of Yorkshire, where employment abounds for persons of all ages, and even for every child who is able to do the least work, it must excite great surprise, that the poor should be so numerous, and the rates so excessive. While we feel most sensibly for the infirmities of old age, and are fully of opinion, that due attention ought to be paid to the distresses of those who are unable to support themselves, we cannot pass over this important subject, without offering a few remarks on the laws presently in force for regulating their support.

Previous to the period when the Reformation took place in England, the poor were supported at the monasteries, and other houses of the irregular clergy, it being then understood, that this was one of the purposes for which tythes were paid to these houses; and after the suppression of the monasteries in 1543, great clamours ensued over the whole kingdom, in consequence of this support being withdrawn. The poor continued in a deplorable state till the 43d year of Queen Elizabeth's reign, when the laws for regulating their support were first enacted, and whatever were the motives which operated upon the minds of our legislators to enact such laws, experience has proved, that the salutary consequences which they expected from them, have been totally unfounded.

The Chancellor of the Exchequer in his speech, February 12. 1796, when Mr Whitbread moved the second reading of the bill, for regulating the wages of labourers, expressed his sentiments upon the construction of the present laws for supporting the poor as follows.

"That the poors laws of this country, however wise in their original constitution, had contributed to prevent the circulation of labour, and to substitute a system of complicated abuses in room of the evils which they humanely meant to redress, and by engrafting upon a defective plan defective remedies, they produced nothing but confusion and disorder. The laws of settlement prevented the workman from going to that market, where he could dispose of his industry to the greatest advantage, and the capitalist from employing the person who was best qualified to procure him the best returns for his advances. These laws had at once increased the burden of the poor, and taken from the collective resources of the state, to supply wants which their operation had occasioned, and to alleviate a poverty which they tended to perpetuate."

With these sentiments we entirely concur, and cannot but regret their not being followed up with a bill or bills for eradicating the evils so justly complained of. In fact the poors rate is the most unequal tax in Britain. It falls entirely upon the possessors of land and houses, while the trading and moneyed interest of the kingdom, pay nothing but for the houses they occupy. When first established, the commerce and manufactures of England were in their infancy, and consequently, permanent or landed property was considered as the only thing upon which an assessment could be imposed. The circumstances of the country being changed, and the number of the poor greatly increased in consequence of manufactures, it appears fair and reasonable that they should now bear their share of the burden, and not cast it wholly upon the landed or territorial interest of the kingdom.

It is within our knowledge, that the present mode of supporting the poor, has in several parts of the kingdom prevented the introduction of manufactures. The landed interest, from dear bought experience, to prevent the increase of the rates, have absolutely refused to allow manufacturers to settle in their bounds, knowing that their establishment is always accompanied with a long train of public burthens. This, from the iniquitous law for regulating settlements, is entirely within their power, and they cannot be blamed for executing this self defensive measure so long as the present laws for supporting the poor are allowed to remain in force.

But the principle of the poor's law is to impose a tax on the industrious, to be paid to the profligate (*a*). It was not many years after it was passed, when the famous song, containing these lines,

Hang sorrow, cast away care,
The parish is bound to maintain us,

was sung in the streets of almost every city in England; and if we resort to experience, or observation, we will find that this sentiment too generally prevails, and contributes to render the lower ranks more thoughtless and extravagant, in the days of health and strength, than they would otherwise be (*b*).

But is no attention to be paid to the distresses of the poor? Most certainly they are entitled to every mark of attention. We only contend, that this ought to be shown to those who deserve it, and that the burthen of their support ought to fall in an equal manner upon all ranks, in proportion to their abilities. We grant at once that those who from age, disease, or debility, are unable to provide for themselves, ought to be furnished with the means of subsistence by the community with which they are connected; but we presume, that the provident support held out by the present laws, goes much beyond what is necessarily required for these ends, and that while they are in force, the number of the poor will continue to increase. Holding out large funds is the sure way of occasioning an increase, as notwithstanding the rates have increased four-fifths at least since the beginning of this century, the number of the poor, under the flourishing state of commerce, manufactures, and agriculture, have also increased. In Scotland where employment is much scarcer, and wages not half so great, the lower ranks by being temperate and frugal, not only bring up large families, but are seldom a burden upon the parish. We are acquainted with country parishes, the population of which is considerable, and the rental betwixt 5 or L. 6000, while the charge of supporting the poor does not exceed L. 60, a considerable part of which is collected at the church door on Sundays, in the way of voluntary charity, and administered by the elders or the kirk session. In a word, we are decidedly of opinion, that the present laws for supporting the poor are founded upon erroneous principles, being not only distressing to the public, but detrimental to industry, and contrary to sound morality, and real religion.

But how is the matter to be mended? how is industry to be encouraged among the lower ranks, the indigent and distressed *pauper* supported, and the burden sustained in an equal way by those capable of bearing it? We answer by going to the bottom of the evil; by repealing the present poor laws, and enacting others more agreeable to the situation of the country; by annihilating the iniquitous law for regulating settlements, and allowing every man to settle where he can find work; by making public support, not a matter of right, but of favour, which may be with-held if the object is undeserving. These things would contribute to amend the dispositions of the lower ranks, would convince them that sobriety, regularity and temperance were the qualifications which would insure them relief, when old age or debility required public assistance, and the practice of those moral qualities would necessarily decrease the number of those who stood in need of such relief.

Perhaps the best mode of supporting the distressed, would be a law obliging every householder to contribute a certain part of his income toward the support of those who stood in need of public relief; the sum to be optional, and the contributor when in distress to draw from the fund in proportion to his monthly, quarterly, or annual payment: To this fund might be added a permanent tax upon landed property, say L. 5 per cent. upon rents, in lieu of the present rates, as there is no reason why the possessors of land should get entirely free of a burden which has affected them for near two centuries. Our object is to prevent an increase of the rates, and to throw the charge of supporting the poor upon the public at large, not to emancipate landed property altogether. This plan, upon the whole, is something similar to those of the friendly societies, (which cannot be too much encouraged,) and if established in every parish, and the funds administered by a committee of contributors annually chosen, would prevent these peculations so grievously complained of under the present system, and in a great measure, put public charity or assistance on its proper basis. We throw out this hint, forbearing to enlarge upon it, under the hope it will be taken up by others more versant in such affairs.

Mr Stockdale at Knaresborough, a gentleman of great intelligence, and much verſant in buſineſs of this nature, has furniſhed us with the following information concerning the adminiſtration of the poor laws.

In Eaſter week, overſeers of the poor are generally nominated from the moſt ſubſtantial part of the townſhip by two juſtices of peace, to ſerve for one year, whoſe buſineſs it is to provide books for their accounts, ſettle theſe of the preceding overſeers, lay a pound rate for the maintainance of the aged and infirm, as well as infant poor within their reſpective townſhips, by ſetting them to ſuch work as they can perform, and theſe powers are in purſuance of two acts of parliament, viz. 43d Eliz. ch. 2d and 17th Geo. 2d ch. 38th.

All impotent poor of whatever age or deſcription, are entitled to parochial charity in the place where they are then reſident, until the laſt place of their legal ſettlement be found; and then on complaint of the churchwardens and overſeers to the juſtices of the peace, they can obtain an order to remove the paupers to ſuch their place of ſettlement; and if the places to which they are ſent are diſſatisfied, and think they ought not to be ſaddled with them, they may appeal to the next quarter ſeſſions, whoſe determination is generally final, but is ſubject to the reviſal and reverſal of the Court of King's Bench; but the paupers muſt be maintained by the inhabitants of the place, where the juſtices ſent them to, till ſuch final determination.

A pauper may come into any pariſh, but he cannot gain a ſettlement there by ſuch intruſion, for he may be taken before two magiſtrates and examined as to his ſettlement, and then removed; but he may gain a ſettlement by renting L. 10 a year; continuing forty days in the pariſh after giving notice thereof in the church; by ſerving an apprenticeſhip to ſome occupation or trade; by hireing for a year; by paying parifh rates, or ferving as a parifh officer; or by coming into the place with a certificate, figned by the church wardens and overfeers of the poor of any other place, acknowledging he belongs to them, and they will receive him back when chargeable; but this certificate muft be allowed by two juftices.

NOTES on Sect. 4.

(*a*) No attention is paid to their morals. Their drunkennefs and profligacy is connived at, or rather encouraged. Vice is rather in efteem, than held in deteftation. Hence their earnings, in profperous times, are fquandered in debaucheries, inftead of being laid up againft the day of adverfity. Their avowed refource is the never-failing poor's rate. *A Freeholder.*

(*b*) I believe there is much truth in thefe obfervations. *W. D.*

I am perfectly in this opinion; for in the little village where I live, the poor are treated with the utmoft kindnefs and humanity; I know of no inftance where they are become more expenfive, in proportion to times paft. Ride a horfe with a flack bridle, and he will ftumble lefs; he will depend upon his own efforts. So it is with the lower order of mankind: the more bountiful we are, the more heedlefs and extravagant they are. I fpeak of the haughty and infolent; the aged and helplefs will, I truft, ever meet with tendernefs and compaffionate affiftance from their fellow-creatures. *A Yorkfhire Farmer.*

Sect. 5.—*Leases.*

The greatest part of the land in this district is not occupied under the guarantee of a lease, the occupiers being generally bound to remove upon a warning of six months. Where leases are granted, their duration is from 3 to 21 years; but three-fourths of the land is possessed from year to year, and this practice, which to us seems destructive of good farming, is upon the increase, although the Duke of Norfolk (*a*) and several other proprietors, much to their honour and profit, act otherwise (*b*). The duty we owe to the public, from the office entrusted to us, renders it necessary that we should describe the ruinous consequences accompanying the want of leases, and how absurd it is to expect that the ground will be improved by persons who may be turned out of their possessions, whenever the proprietor, or more properly speaking the steward appointed to manage his estate is disposed, by caprice, whim, enmity, or interested motives, to give them a warning of removal (*c*).

That celebrated agricultural writer, Arthur Young, in his Political Arithmetic, published *twenty years ago*, has said that, "the improvements which have taken place in England, have been almost owing to the custom of granting leases, and that in those counties where it is unusual to grant them, agriculture continues much inferior to what it is to be found where they are usual." If this doctrine be admitted, (and in our opinion it is founded upon principles that cannot be disputed,) the general custom of not granting leases in the district we are now treating of, must deserve reprehension; and if we are to judge of its husbandry by the rule here laid down, we would be under the necessity of declaring, that however flourishing the country may be, and however much it may be improved in every branch of its agriculture, still if leases had been granted, and a security thereby offered to the farmer for enjoying the fruits of his labour, these improvements would have increased; and consequently the interest not only of the public, but also of the proprietors themselves, would have been materially promoted (*d*). This is an important subject, and well deserves the attention of every landed gentleman in the kingdom (*e*).

Before a farm can be put in proper order, a considerable time must elapse, and much money must be expended. The fruits of improvements are not gained all at once, and a number of years are required to accomplish the best digested plan. Suppose, for instance, a person entering to a farm that was worn out and exhausted by long and successive tillage, and that he wishes to refresh the land by laying it down in grass; it will be six years at least before he can go over it all with fallow, and unless he sow it down clean, he is neither doing the land nor himself justice (*f*). If he continues it in grass five or six years more, which is little enough time for ground so exhausted, it will be found that near twenty years must take place before he receive the reward of his improved cultivation; and to receive this reward he has a claim both from his superior management, and as an incitement to his future industry: but what security has he for this reward, or what incentive has he to industry, if he sits upon the premises by virtue of an annual lease (*g*). In the midst of his career he may be interrupted by a *six months warning*, and the toil of his hands, and the fruits of his improvements, go to another. These are not imaginary apprehensions, but are founded upon real and solid principles; and which will operate less or more upon every farmer, according to his ſituation and circumſtances (*b*).

Many caſes of a ſimilar nature might be put, but from the above we hope it will appear, that before any ſubſtantial improvements can be expected from the farmer, he muſt have the ſecurity of a leaſe, for affording him time to reap the fruits of theſe improvements. There is, in the courſe of farming, as much often laid out in one year, as many ſucceeding crops can repay*; in this caſe, where the farmer has a leaſe, he looks to a future period for being reimburſed: if he has none, can it ever be expected that any man of common ſenſe will throw away his money by improving another perſon's eſtate, and caſt himſelf upon the mercy and diſcretion of his landlord for time and opportunity to gain it back again? The farmer who would do this, is not guided by the ſame principles that influence the reſt of mankind.

The more a farm is improved, the greater the quantity of manure laid upon it, the cleaner the fields, the richer

* We ſhall give one inſtance to corroborate what is here ſaid. A farmer of our acquaintance had an acre of rich moſſy meadow ground, which was totally unfit for ploughing, and could ſcarce carry the weight of a beaſt in the drieſt ſummer months. In order to make it crop with the reſt of the field, he drained it completely; and, as from the ſtrength of the roots of the herbage it would not plough to advantage, he digged the whole of it this ſeaſon with the ſpade, and propoſes to lime it after the firſt crop, when it is expected the ground will be conſolidated. The expences were,

Caſting drains - - - - -	L. 4	15	0
Gathering ſtones, driving them, and filling up the drains	5	18	0
Digging the ground, which, from the ſtrength of the roots, was a ſevere operation - - -	4	10	0
Total expence	L. 15	3	0

Beſides the expence of lime, which will be L. 6 more.

Query, Would he have improved this meadow without a leaſe?

the paſtures and meadows, the completer the fences, and the more convenient the buildings and offices, are all circumſtances that may operate againſt the farmer who has no leaſe, and be the means of alluring a covetous neighbour to attempt wreſting his poſſeſſion from him, or may be uſed as arguments by a deſigning ſteward for raiſing his rent. Such being the caſe, every conſiderate man is deterred from expending a halfpenny more than he is neceſſarily obliged to do; and therefore it follows, that the withholding leaſes is a real and certain obſtacle in the way of farther improvements (*i*).

We might alſo mention arguments of another kind for granting leaſes, which, however contemptuouſly they may be viewed by others, have great weight with us. The farmer who ſits without a leaſe, has not the privilege of thinking and acting for himſelf;* it is needleſs to bring forward arguments in ſupport of this propoſition, for it cannot be contradicted. We have often heard it ſaid, that the liberty enjoyed by the farmer, and the ſecurity afforded by the conſtitution to his property, were the principle cauſes why agriculture flouriſhed more in this iſland than in other nations. We beg leave to inquire, where is the liberty enjoyed by the farmer who ſits without a leaſe? his words and actions are under the moſt abſolute ſubjection to another, who carries along with him a never failing argument upon all occaſions. Let the abject ſituation of ſuch a man, placed under a capricious landlord, be conſidered, his beſt actions may be miſinterpreted; he is expoſed to every indignity without daring to complain: or if the

* We were informed the tenants on an eſtate in the Weſt Riding had got warnings of removal, merely becauſe *they had turned Methodiſts*. There are not many landlords that find fault with their tenants for being religious (*k*). This inſtance is only given to ſhew upon what trivial grounds removals are made.

ſpirit of a man gets up in him, what ſecurity does the conſtitution afford to his ſituation ?† If he has made improvements, the fruits of them are wreſted from him by an arbitrary removal (*l*). Another farm cannot always be got, and he may be turned upon the wide world without the hopes of redreſs. A prudent man will reflect upon theſe things, and if he is ſo critically ſituated, will often rather part with his natural rights than expoſe himſelf to miſery: he may have a numerous family; his farm may be doing well with him; he may have contracted an affection for his *natale ſolum*, and be uncertain, if he makes a change, how he is next to be put up. The picture may be ſtill higher coloured; but from the above we contend, that the want of a leaſe precludes the farmer from acting as a free agent, and renders his property inſecure and precarious (*m*).

(†) The cuſtom of the country in allowing what is called tillage, and half-tillage, to the out-going farmer, is no reimburſement for any improvement he may have made. The time of entry is at Candlemas, and the incoming tenant enters to the wheat that is ſown, and to the labour done upon the farm by his predeceſſor; for theſe things, as well as the manure laid on, and the graſs ſeeds ſown the preceding year, he is allowed; but as for money expended upon buildings, incloſures, drains, or other ſubſtantial improvements, which add to the permanent value of the property, he receives no reimburſement at all.

NOTES on Sect. 5.

(*a*) His maxim is, "let them thrive;"—too many adopt the contrary maxim, "keep them down."

A Freeholder.

Extract of a letter from a gentleman near Sheffield to Mr Brown.

"The whole of the extensive estates of his Grace the Duke of Norfolk, in this neighbourhood, are, generally speaking, let upon leases of 21, 42, 63, and 99 years. For farms the first is the usual time granted; but where any extensive erections have been, or are intended to be made upon the premises, there is no difficulty of procuring a lease for any of the longer terms conditionally, that in *proportion to its length* there is a certain increase of they early rent put upon the property. Perhaps no stronger instance need be adduced, in favour of leases for a term of years being granted to the occupiers of landed property throughout the kingdoms, than the beneficial effects which result to the community in this neighbourhood, from this liberal system being pursued by his Grace's agents."

(*b*) I sincerely wish every proprietor thus sensible of his own interest, and that of his country; for, without a lease, the most useful member of society is degraded to a slave. He is not only debarred from managing his farm in the spirited manner he would wish, but, if he is near his landlord, he is afraid of either riding on a good horse, or putting on a good coat. In short, he must neither think nor act for himself, but be for ever subject to the whim and caprice of those he lives under. *There are, no doubt, many exceptions from this, myself among the rest;* but it is *too often* the case. Happy are they, (without a lease), whose landlords are of too liberal a disposition, even to suffer them to feel the want of one. *A Yorkshire Farmer.*

This is the greatest obstacle to improvements, and every well-wisher to his country ought to exert himself in helping to remove it. It can never be expected that husbandry will be brought to any tolerable degree of perfection, unless the occupier has the security of a lease. It is true we have a few gentlemen, who, much to their honour and intereſt, have acted upon ſuch principles as entitle them to confidence; but the time may come, when, by the courſe of Providence, theſe reſpectable characters may be removed, and we may be thrown into the hands of perſons who will take advantage of our induſtry. Adieu then to future experiments and future improvements—for the beſt farmers would in ſuch caſes be the greateſt ſufferers.

A Farmer.

(*c*) The tenantry are very much plagued by *attorney ſtewards*, &c. who muſt have buſineſs, or otherwiſe make it.

A Freeholder.

(*d*) So ſtrong a caſe doth not uſually occur. If a land-owner was fully ſatisfied that the tenant was willing and able to do all theſe things, he would act wiſely in granting him a ſuitable leaſe.

T. York, Eſq.

Anſwer.—The caſe occurs every day. It is the landlord's fault if he does not procure a tenant *able* to do what is neceſſary; and unleſs he give him a leaſe, he cannot expect him to be *willing*.

R. B.

(*e*) Let them but grant leaſes, and they will moſt aſſuredly experience the heart-felt ſatisfaction of beholding their eſtates improved, and their tenants happy. *A Yorkſhire Farmer.*

(*f*) The juſtneſs of the reaſoning here uſed, appears inconteſtible. *Anonymous.*

(*e*) Some landlords conſider their tenants as merely ſtewards or bailiffs, and raiſe or lower their rents according to the price of corn. A more vague criterion cannot be adopted than this; for, on many farms, a high price of corn is the effect of an unfavourable ſeaſon. Now, it is well known, that high price is ſeldom a compenſation for bad crops, and the farmers rich years, are thoſe in which a moderate price of corn is the effect of an abundant crop.

A Favourable Showery Season.

Barley, 5 quarters, at 26s. - - - -	L. 6 10 0
More straw per acre - - - - -	0 10 0
	L. 7 0 0
	W. P.

An Unfavourable Dry Season.

Barley per acre, 3 quarters, at 40s. -	L. 6 0 0
Balance against high price - - -	1 0 0
	L. 7 0 0

(*b*) In my opinion, these are unanswerable arguments in respect of leases. *Mr Cioley.*

(*i*) The reasons here assigned, in favour of leases, are so powerful and well founded, that it is hoped every unprejudiced liberal minded proprietor will see it is his best interest to grant such security to the occupiers of his land. *A Farmer.*

(*k*) Many gentlemen in the county of Essex, to the distinguished *honour* both of their *heads* and their *hearts*, have dismissed their tenants for being Dissenters, though possessed of every other requisite of character and conduct.—Curious proofs of an enlightened age! *Anonymous.*

He who deserts an established religious rule, that aids him in the performance of every moral duty, for that misguided zeal, which wanders in caprice and error, is not the person in whom confidence can be satisfactorily placed. I do not by this observation intend to oppose the practice of granting leases;—on the contrary, I think it founded on equity; but the tenant to whom they are granted, should possess *stability* in religion, as it is the most powerful incentive to the observance of moral obligation. *W. Fox.*

Answer.—The above observation is weak, illiberal, and absurd. The writer supposes no moral duty can be performed without the pale of the Church of England; and in fact, goes the length of denying *fire* and *water* to any person who deserts the establishment. *R. B.*

(*l*) Yet a compensation might be settled by law, including every possible improvement as a part of stock in trade, to be paid to the quitting tenant. I confess I would not take a farm on *lease* and tie myself to pay high rent and *encreasing taxes*, whatever *may happen* during the term, to raise the out payments, or abate the price of products. *W. P.*

(*m*) But a long lease renders the value of the property very precarious, and dependent entirely on the good will of the farmer. There will be loop holes in the best contrived covenants, which a knave may take advantage of; and, if he can pay his rent, the landlord must go to law, the issue of which is precarious.

Messrs S. P. & M.

Answer.—How can the landlord's property be injured by the independency of the tenant? It might with much greater propriety be urged, that the independency of the tenant will enable him to cultivate his fields in a superior manner than he could do, if his condition was different. If the tenant implements the covenants contained in his lease, where will the loop holes be, which will give him advantage over his landlord? If he does not, a summary process can easily be brought to compel him. *R. B.*

SECT. 6.—*Covenants in Leases.*

THE covenants which subsist in the agreements for land betwixt landlord and tenant, are many and various. We were favoured with copies of several of these agreements, and had opportunities of seeing others in the hands of the possessors. We shall give an abstract of the clauses contained in some of them now before us.

In one of these, the covenants are as follow:——The landlord sets the grounds for 10 years, and gives entry to the land on the 2d day of February, and to the houses upon the 12th of May: the rent to be paid in equal portions, at the first term of Whitsuntide and Martinmas thereafter. Reserves the liberty of hunting and fishing on the premises, and the property of all mines and quarries, and the iron ore, coal, lead, or other minerals contained in them. Reserves liberty to go into the inclosures to cut and dig trees of all kinds, with access to carry them off. The tenant obliges himself to pay all taxes, as well parliamentary, as other ones already imposed, *or to be imposed during the currency of the lease*, without desalcation from the rent. Obliges himself also to eat all his *hay* and straw upon the premises, and to dung a part of his meadow ground every year. Agrees not to plough any of his old pasture under a penalty of L. 10 per acre, nor to have above one fourth of his farm under the plough at one time (*a*).

The lease also contains a great many clauses, about attending courts, repairing fences, grinding malt and corn, &c. &c. &c. which it is unnecessary to mention.

In another we observe the following conditions:

Restricted from ploughing any of the meadow or pasture land.

Obliged to fallow the third part of the tillage land annually, and to lay two chalders of lime upon every statute acre.

To pay all parliamentary and parochial taxes at present existing, or that may be laid on during the continuance of the lease.

To keep up all fences roads, bridges, &c. upon the farm.

To pay the rent within twenty days after it becomes due, under forfeiture of the lease.

To pay a penalty of L. 10 for every acre not managed agreeably to the covenants, over and above the rent.

Conditions of a third lease:

Entry to the farm at Candlemas.

Rent payable at Whitsuntide and Martinmas thereafter.

No *hay* or straw to be sold.

No meadow or pasture to be ploughed without consent of the proprietor.

When land is sown down for grass, to be done with 12 *bushels of fine hay seeds*, and 4 lbs. of Dutch white clover per acre.

Tenant removeable at 6 months warning.

In other leases we saw, the tenants were expressly prohibited from breaking up all grass lands that have lain 6 years, which renders the situation of the pasture and meadow fields as immutable as the laws of Media and Persia were of old. In short, the very nature of most of the subsisting covenants are destructive to improvements; and, as it was well said by Mr Potter at Tadcaster—" A good farmer will manage much better wanting them, and as for a bad farmer, they never will mend him."

The following is copied from a paper given us, and is the fubftance of the covenants entered into on the eftate of that benevolent and public fpirited nobleman, Earl Fitzwilliam.

The tenant covenants to keep all the buildings and fences in repair; to pay all parliamentary and parifh taxes; not to plough up grafs land without confent of he landlord; not to take more than 3 crops of corn before a fallow; to lay 12 cart-loads of dung upon every acre fo fallowed; not to fell any hay, ftraw, or other fodder from off the premifes, but eat and confume the fame thereupon; to fpread all the manure arifing from the premifes upon fome part thereof, and leave the laft year's manure thereupon. The landlord covenants to allow the tenant, on quitting his farm, which is by the cuftom of the country at Candlemas, what two indifferent perfons fhall deem reafonable for what is generally called full tillage, and half tillage, being for the rent and affeffments of his fallow ground, the ploughing and managing the fame; the lime, manure, or other tillage laid thereon; the feed fown thereupon; the fowing and harrowing thereof; alfo for the fowing, harrowing, manuring, and managing any turnip fallow, which he may leave unfown; alfo for any clover feed fown on the premifes, and harrowing and rolling in of fuch feed; and for every other matter and thing done and performed in a hufbandry-like manner on fuch fallow lands, in the two laft years of the terms; alfo for the laft year's manure left upon the premifes; and for any manure and tillage laid upon the grafs land.

The primary error of the Yorkfhire hufbandry confifts in not giving the tenant a fecurity of poffeffion for a reafonable time; and the fecond, and no lefs important error, arifes from the reftrictions impofed during the time he occupies his farm, which prevents him from changing his management, or of adapting his crops to the nature of the foil he poffeffes. Agriculture is a living fcience, which is progreffively improving, confequently what may be efteemed a good courfe of cropping at one time, may, from experience and obfervation, be afterwards found defective and erroneous.

That particular covenants in a leafe are obftacles to improvements cannot be difputed; for the very nature of a covenant fuppofes that the practice to be regulated by it had arrived at its *ne plus ultra*, and could not be mended. Thefe covenants or reftrictions fubfift more or lefs in every leafe we heard of; and the fhorter the leafe the more numerous they are. In annual leafes there appears an abfolute neceffity for them; as the farmer, from having no certain profpect of enjoying his poffeffion, would otherwife be tempted to difregard every branch of good hufbandry.

It will hardly be alledged, in defence of this practice, that agriculture has already arrived at its utmoft pitch of perfection, and that improvements in that art can be carried no farther. We will not fuppofe that any perfon acquainted with the fubject will offer fuch defences. The very appointment of the Honourable Board, for whofe confideration this is drawn up, is a public teftimony that the practice of hufbandry may ftill be improved. But how is this to be done if the farmer, who is the firft wheel of the agricultural machine, be reftricted in his management? If the crops he is to fow be marked out by the drawer of his leafe, how are more approved rotations to be introduced? The fact is, that all good farming is local, and muft in a great meafure be regulated by the foil and the weather. It is therefore abfurd to lay down in a leafe particular rules for a number of years practice; as, from circumstances, many fields are often both richer and cleaner after carrying 5 or 6 crops, than others are after two; consequently, without leaving these things to the wisdom and judgment of the farmer, the ground can never be properly cultivated, nor made to produce its greatest value (*b*).

Restrictions in a lease necessarily suppose that the framer of them possessed more knowledge of farming than he whose operations are thus to be directed (*c*). We leave the public to judge whether this can actually be the case or not. Leases are often copied from one generation to another, without paying any attention to more recent improvements. How is it possible for an attorney, or his clerk, to lay down rules for the farmer's direction? Allowing it is the steward, or even the proprietor himself, that dictates these rules, we are warranted to say it is naturally impossible they can be wisely and judiciously framed (*d*). Laying aside the consideration of their fettering the farmer's mind, and clogging his operations, such restrictions or rules may, from alteration in markets, be unprofitable; and from the vicissitudes of seasons improper to be executed.

Every farmer knows from experience, that the proper manner of cultivating land is only to be learned from an intimate acquaintance with the nature of its soil, and that what is very good management upon one farm, is often very bad upon another. The covenants suppose all to be alike, that grass is of equal benefit on all lands, and that the same quantity of lime should be administered to a light loam as to a strong clay. Besides, in framing these covenants, it is taken for granted that a person from cursory view, is at once able to determine upon the best mode of management for the endurance of a whole lease; or, in other words, that his judgement is equal to that of the whole tenantry of an estate. In short, restrictions are inimical to all good husbandry. They sink the farmer into a state of insignificance. They contract his mind, and lock up his ideas from searching after new schemes, which is the only method by which improvements can ever be found out; and therefore it follows, that a continuation of covenants is highly detrimental not only to the public good, but even to the interest of the proprietor himself, by lessening the rent that a superior cultivation, arising from a spirit of improvement, would be able to pay (*e*).

We are ready to admit that general rules of management are very proper in leases, such as, to keep the farm in good order, to consume all the straw raised upon it, and to sell no dung. These restrictions we will allow; and every good farmer will follow them whether he is bound to do so or not. Nay, we will go farther—If leases of a proper duration were granted, it is very reasonable that the property of the landlord should be protected by restricting clauses for the 3 years previous to their expiration. But after all, it will be found that no clause can be inserted, besides the general ones already mentioned, that will serve to enhance the value of the land, except obliging the farmer, to leave a proportional quantity of such land in grass at the expiration of the lease, and specifying the manner in which that land is to be sown down. Other clauses serve only to distress the farmer, but will never promote the interest of the landlord (*f*).

NOTES on Sect. 6.

(a) Some of these covenants are of such a pernicious nature, that no man who wishes well either to himself or the public, would undertake to perform them, unless in particular cases and situations. *T. H.*

Rather than enter to a farm upon such conditions, I would sell off and go to America; and I sincerely wish every farmer in the kingdom to be of my way of thinking in this respect.

A Yorkshire Farmer.

(b) If the covenants are framed to secure the practice of the most approved course of husbandry, they do not bar the tenant from improvements. It is in the superior cultivation of these crops, that his skill is to appear; and he is rather aided, in my opinion, than fettered, by having good rules to govern him.

W. Fox.

Answer.—There is a good deal of sophistry in the above observation. Mr Fox first sets up a man of straw, *i. e.* he supposes the covenants to be so framed, as to secure the most approved course of husbandry, and then argues upon their utility. It almost makes one lose patience to hear of rules being laid down for governing the farmer. Agriculture is a living science, which is progressively improving, consequently what may be esteemed a *good* rule one year, may, from experiment and observation, afterwards be found erroneous and defective. *R. B.*

(c) I agree in this opinion, and conceive that it is worthy the consideration of the proprietors in general, whether covenants in leases, to compel the occupiers to a certain routine of crops, (be the seasons what they may,) until the last three or four is for their benefit. Injurious to the tenant it certainly is, and in my humble opinion, highly improper. *W. D.*

(d) Comparisons are invidious; however, the estates which are under the immediate inspection and controul of their owners, are usually distinguishable by the superiority of their cultivation; and the estates of absentees, and others who assign full power of management to tenants, without covenants or restrictions, are generally remarkable for their slovenly and impoverished conditions. Most improvements have been made and introduced by men of education and fortune; by them agricultural knowledge is diffused, by precept, by example, by publications, and conversation, and, in fine, by such wholesome covenants as the tempers of their tenants will bear. It appeared by a work published some years ago, called the "Northern Tour," that many well informed gentlemen resided at that time upon their estates in Yorkshire, who practised a very improved system of agriculture; if such should be induced to resign the study of it to illiterate farmers, the knowledge would immediately fall into decline; and if, by chance, an useful improvement should be discovered in one county, ages might pass away before it would be introduced into another.

T. York, Esq.

Answer.—We have noticed, in a cursory manner, a few of Mr York's observations; but the one now before us requires a stricter examination: It goes the length of asserting, that agricultural improvements can only be advantageously executed under the controul and inspection of landed proprietors, and, if well founded, would, in a great measure, overturn what is said in this Survey upon the important articles of leases and covenants.

It is fashionable for landed gentlemen to attribute the merit of agricultural improvements to their own body, and to accuse the farmers of ignorance, obstinacy, and inattention to useful discoveries. We might here inquire, How it comes about, that landed gentlemen claim a superiority over the operative farmer in rural knowledge, while they tacitly allow people of every other profession to possess greater knowledge in the various occupations which they practice? That many proprietors have introduced and encouraged improvements, we are not to deny; but to suppose that *most* improvements have been introduced by their means, or that their management must necessarily be superior to that of the actual farmer, is extravagant and absurd.

Agriculture, though apparently a simple science, is only to be learned by a diligent attention to practices and circumstances, which, in the language of the world, are below the notice of men of fashion and property. Are we to expect, that persons of this description are to rise with the sun, and to toil till he goes down, in superintending the different processes of farm management? This is never to be expected; their education and habits of life rendering them unfit for such sedulity; and the consequences attending the want of it have been evident, whenever proprietors stepped out of their own line to farm any considerable part of their estates. While they farm for convenience or amusement, their intention is laudable and innocent; but it would only be paying them an unmeaning compliment to say, that improvements can be more judiciously introduced, or more frugally executed, under their inspection, than under the direction of professional men, whose subsistence must necessarily depend upon the success of their exertions.

Mr York's first assertion is, "That estates under the immediate inspection and controul of the owners, are usually distinguished for the superiority of their cultivation; and the estates of absentees, and others who assign full liberty of management to tenants, without covenants or restrictions, are generally remarkable for their slovenly and impoverished condition." Without inquiring into the justice of this comparison, (which, at the same time, we suspect to be erroneous) we beg leave to remark, that Mr Y. takes up the subject more according to the present narrow and limited condition of the tenantry, than if their situation was meliorated and improved. During our survey, we heard of few, or rather of no, tenants who were allowed discretionary management, and the portion of our report upon which he founds this observation, was wrote under that impression. If we were wrong in saying, that the majority of tenants had no leases, and that they possessed their farms under what we thought ruinous and destructive covenants, our errors, in these respects, ought to have been pointed out; but as the account which we gave of the nature of the connexion betwixt landlord and tenant is not contradicted, we are warranted to suppose it is fairly stated. Again, with regard to the pernicious tendency of covenants; that they fettered the mind, and clogged the operations, of the farmer, were improper in many cases, and impracticable in others, he does not offer a single argument in refutation of our doctrines, but contents himself with running a comparison which, at this distance, it is morally impossible for us to follow out.

In the West Riding, the fundamental error that takes place in the management of estates, arises from considering the tenant as possessing little more knowledge than the horse he drives, and as destitute of abilities to manage the ground in a proper manner. Hence proceeds the numberless covenants which are to be found in every agreement for land, which only fetter the tenant, and prove detrimental to the public intereſt, without being of uſe or advantage to the landlord. It can hardly be ſuppoſed, that perſons of the moſt liberal principles will take land under ſuch arbitrary conditions, nor are we to expect that the operations of those who do ſubmit to them, are to be carried on with equal vigour and ſpirit, as if the management was left to their own knowledge and judgment.

Mr York appears to make up his mind in conformity to the preſent ſyſtem, without attending to what might be done by the tenantry, if that ſyſtem was altered. This is not doing juſtice to the ſurveyors; for the ſcope of their arguments go to ſhow, that if leaſes of a proper duration were granted, and freedom of management allowed, the exertions of the tenant would increaſe, and conſequently the public good would be promoted. It can never be known what a farmer will do, unleſs the ſecurity of a leaſe is previouſly granted him, and without giving him that ſecurity, it is unreaſonable to expect he will improve the ground he poſſeſſes.

There cannot be ſtronger proofs exhibited of the happy conſequences reſulting from free and open leaſes, than the great and ſubſtantial improvements executed in all the cultivated counties of Scotland, which no man in his ſenſes would have undertaken without that ſecurity. Excellent farm houſes and offices have in many places, been erected; open fields have been incloſed; wet lands have been drained; and lands, formerly unproductive waſtes, have been brought into a comparative high ſtate of culture. Whereas, if Mr York's ſentiments were juſt, the country, inſtead of being improved thereby, would have been reduced to utter deſtruction.

And pray, has the value of the landlord's property been leſſened by theſe leaſes? According to Mr Y. a dreadful havoc might be expected at their concluſion; the buildings and fences would be in ruins; the land exhauſted; and a great fall of rent the unavoidable conſequence. No ſuch things have, however, happened; the tenant, knowing he cannot ſcourge the ground, except for two or three of the concluding crops, without ſcourging himſelf, naturally does every thing to render the land fertile and productive, and, inſtead of rentals decreaſing, they have increaſed in a two-fold degree greater than in the beſt cultivated counties of England.

Another of Mr York's aſſertions is, that if gentlemen of land-property reſign the ſtudy of agriculture to illiterate farmers, as he is pleafed to ftile them, the knowledge of that fcience would fall into decline, and ufeful difcoveries would not be diffeminated. We are at a lofs to difcover where we recommended fuch a refignation to the landed proprietor, and apprehend the field of agriculture is wide enough for both. It furely does not follow, becaufe we recommended the granting of leafes free from ufelefs and pernicious reftrictions, that the proprietor was thereby to be denied liberty of farming any part of his eftate, or the whole, if he thought proper; and it certainly will be granted, that any fcheme which contributed to enlighten the farmers, would of courfe forward the circulation of agricultural improvements with double rapidity.

According to Mr York, the landed gentlemen of the Weft Riding poffefs infinitely more knowledge of hufbandry than the actual farmer. They promote improvements by precept, by example, by publications, by their converfation, and ftill more by the *wholefome covenants* they prefcribe for their tenants; which are adminiftered, not according to the nature of the foil they poffefs, as ought to have been the cafe if covenants were ufeful, but according "*as the temper of their tenants will bear them.*" Without meaning the fmalleft difrefpect to the proprietors of the Weft Riding, we beg leave to fay, that if what we have ftated be juft, the merits of the caufe would be directly reverfed. In a word, the bounties of Nature are always difperfed with an equitable hand; and, while the fee-fimple of the ground is conferred upon one man, the talents and abilities for rendering that ground fertile and productive, are generally beftowed upon another. *R. B.*

(*c*) It muft be abfurd beyond all doubt, to fuppofe fuch a thing. If the farmer is a fenfible man, let him have liberty without reftraint, and if he does well for himfelf, he moft certainly will do fo for his landlord: If he is otherwife, he is not fit for a farmer; for it is my opinion, to be a proficient in agriculture, and in the knowledge of ftock, requires as much ftudy and application as any other fcience. A good education is a very neceffary ingredient in making a good farmer, and the want of it is a very great obftruction to improvements, by contracting the ideas, and rendering the faculties incompetent to the contemplation of theoretical knowledge. But all the education, and all the experience that can be united with the greateft ablities, can never be fhown to perfection in a farmer, without a leafe.

A Yorkfhire Farmer.

(*f*) These are spirited, but very just remarks. *Mr Culley.*

If the proprietors of land were sure of always getting tenants that would act properly, there would be no need of restricting covenants; but this is not always the case, and there are many instances of estates being much injured by exhausting crops where tenants were not properly restricted. That many covenants are useless or hurtful, I readily admit; but covenants may be so framed, that a tenant shall have ample liberty to take such crops as he shall think proper, and to propose such modes as shall benefit himself, without injuring his landlord. *Mr Bailey.*

CHAPTER V.

IMPLEMENTS.

IN no practice are the farmers of the West Riding more defective than in the construction and management of their ploughs and wheel-carriages. These are material articles of rural œconomy, and are generally most perfect where the best husbandry prevails.

The Rotherham plough has been heard of over the whole island, and was invented by Mr Joseph Foljambe, of Eastwood, in this Riding, about seventy years ago. Mr Foljambe got a patent for this plough, which he afterwards sold to Mr Staneforth of Firbeck, who at first gave the liberty of using it to the farmers for 2s. 6d. each. Mr Staneforth afterwards attempting to raise this premium to 7s. or 7s. 6d. the validity of the patent was combated and set aside, on the ground of its not being a new invented plough, but only a plough improved. It does not fall within our province to investigate the causes of this decision; but certainly if Mr Foljambe deserved to have a patent right in the first instance, for his invention, that right was in no shape affected from the circumstance of ploughs being constructed long before his time.

The dimensions and construction of this plough will appear sufficiently evident from the following draught and description:

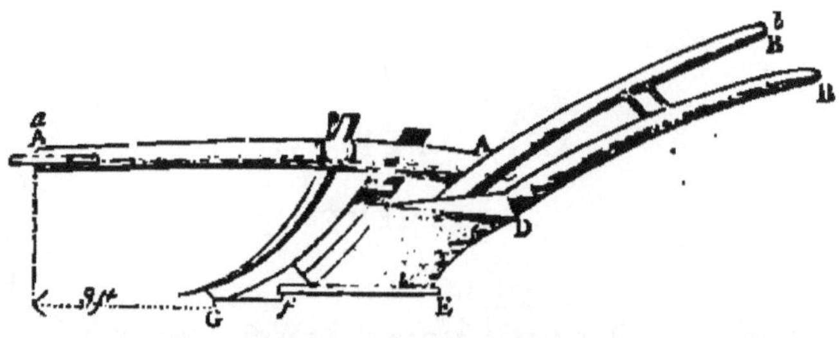

Hock with teeth, to admit of more land being given to the plough, and *vice versa*.

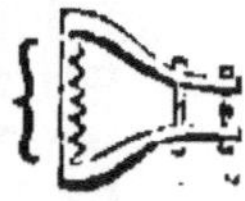

Dimensions of the Rotherham Plough.

	ft.	inch.	
From the end of stilt B to point of the Share G -	7	4	whole length.
From the end of Beam A to ditto of ditto G - -	3	0	
Length of the beam A A -	6	0	
Width of the head in the widest part D - -	1	4	
Ditto of Ditto at E -	0	9	bottom working surface.
Ditto of share behind the wing at *f*	0	$3\frac{1}{2}$	
Length of surface on which the plough touches the ground E G - -	2	$10\frac{1}{2}$	
Height from ground to top of beam where coulter goes through	1	8	
Width between stilts at the end B B - -	2	6	
Height of ditto from the ground	1	11	

Weight of wood and iron work, about $1\frac{1}{4}$ *Cwt.*

This plough, with a few trifling alterations, is used over the whole district, and from being commonly called the Dutch plough, we are inclined to think must have originally been brought from Holland by Mr Foljambe. The faults of this plough are more owing to the manner it is wrought, than to the principles on which it is constructed; for the horses being in many places yoked in a line renders it necessary to turn the beam considerably to the furrow, in order to give the *plough* what is technically called land. Owing to this erroneous manner of placing the beam, the horses draw in a contrary direction to the share and coulter, which makes the plough go unsteady, and from the difference betwixt the direction of the draught, and the head on which the share is fixed, the force of the resistance must necessarily be increased, and the work imperfectly performed (*a*).

Notwithstanding the necessity of turning the beam towards the furrow is entirely owing to the custom of yoking horses in a line, yet we observed, even when horses were yoked abreast, that the ploughs had all more or less of the same direction. The sock or share is much broader in the point, than those we are accustomed to use, which must make them difficult to work on gravelly soils, and even in clay, when the ground is dry.

The practice prevailing over at least one half of the Riding, of yoking horses in a line, is truly absurd. Horses never work so easy or draw so equal as when yoked a-breast, or in pairs, nor will the work be done well in any other manner; if the ground is in that situation as not to bear a horse on the unploughed part, it is unfit for labouring and ought not to be touched. But this cannot be sustained as a reason for this practice, as we repeatedly saw three horses in a line, sometimes even four, ploughing tender clover leys. The plea of custom and prejudice is well known, and can only be assigned for such an absurd and unprofitable practice.

We are clearly of opinion, that every part of plough-work may be executed by two good horses if they are properly maintained. We speak from what is daily done on our own farms, where land fully as strong as any we saw in Yorkshire is constantly ploughed with two horses, and from any thing we saw during our survey, a deeper furrow is generally taken. There is no question but where land is hard and stiff, so much work cannot be done in a given time as upon lighter soils. But this argument will have the same weight whatever number of horses are yoked; all we contend for is, that two good horses yoked abreast, in a plough properly constructed, are able to plough any ground when it is in a proper situation for being wrought.

It is proper to notice, that owing as we suppose to yoking horses in a line, the work is often very defectively executed. There is hardly a straight ridge to be seen, and the ridges are generally kept too flat, not being sufficiently high for setting off the winter rains. We observed this particularly between Thorn and Snaith, where notwithstanding the land is incumbent on a wet bottom, yet the ridges were narrow and not gathered off the flat. At the same time it gives us pleasure to add, that the land near these places was much neater ploughed than any we saw during our survey.

We often remarked, that the land was ploughed too shallow, which not only occasions the pasture of the plants to be curtailed, but also exposes them to be burned up by drought in one season, and drenched by moisture in another. We would lay it down as a rule never to be departed from, that all land should be ploughed in direct proportion to its depth, and that where the soil will admit, it ought to be done substantially.

The farm carriages are carts and waggons of various dimensions. The carts in general are out of all proportion, being far too narrow, and, what is worse, of great length, which makes them heavy on the shaft-horse when going down hill, and to have the contrary fault in the ascent. They are difficult to unload, when employed in driving out dung or performing any home work; and from the sides folding inwards, instead of casting out to the wheel, hold much less than at first sight they might be thought to do. They are drawn by 2, 3, and 4 horses, and are very unhandy about a farm (*b*). The waggons are both upon broad and narrow wheels; but whatever way they are mounted, they prove in the highest degree destructive to the roads, and, in our opinion, are not of the smallest advantage to the farmer.

About Rotherham and Sheffield, the carts and waggons are of the following dimensions:

Carts with 3 horses, narrow wheels, 7 feet long, 3 feet 6 inches wide, 1 foot 8 inches in depth; weight about 12 cwt.

Waggons with 4 horses, narrow wheels, 12 feet long, 4 feet wide, 1 foot 8 inches in depth; weight about a ton.

We suppose that carts of a shorter construction, and rather wider, with sides throwing out to the wheel, and of a size to be drawn by 2 horses, are preferable to those presently used. If a person will attentively consider the manner in which horses do work in a cart, he will soon be convinced of the impropriety of yoking too many together. We are decidedly of opinion, that the lighter the cart, and the fewer the horses, the more loading will proportionally be carried; at the same time a great saving will be made in the important articles of *tear* and *wear*.

There is another branch of agricultural implements

remaining to be defcribed, that is, the threfhing machine; which, in a public point of view, may be confidered as the greateft practical improvement ever introduced into this ifland.

No part of farm-work caufes fo much lofs and vexation to the farmer, as the procefs of feparating the corn from the ftraw, and various methods have, in different ages, been adopted for accomplifhing this operation. The ancient inhabitants of Afia and Egypt, where agriculture is fuppofed to have had its origin, knew no other method than that of inclofing a fpot in the open air, and fmoothing it with clay rolled hard; this was the threfhing floor. The corn being next fpread in fheaves, oxen were turned in, and kept in motion till the bufinefs was done. "Thou fhalt not muzzle the ox that "treadeth out the corn," Deut. xxv. 4.

If Œlian may be believed, the Greeks were neither fo merciful or cleanly in this circumftance. They befmeared the mouths of the poor animals with dung, to keep them from tafting the corn under their feet. Hift. Animal, l. 4. ch. 25.

Machines were next invented, in different countries, made of planks or beams, ftuck over with flints or hard peggs, to rub the ears between them, others to bruife out the grain by fledges or trail carts

Dicendum et quæ fint duris agreftibus arma
Tribula, traheæque, et iniquo pondere raftri.

The tranflators of Virgil, from Father Ogilvy downwards, have included the flail in this defcription,

The fled, the tumbril, hurdles, and the flail. DRYDEN.

Tribulum, however, was certainly the machine firft defcribed for the fingle purpofe of feparating the grain from the hufk or chaff. At what period of time the flail took place of the former awkward machine, is not known with certainty. President Goguet says, that the Turks and many of the Italians, have not yet adopted it. The barbarous Celts, accustomed to fire and sword, made short work. They burned the straw, and instantly devoured the grain; and it is said this custom continues in some parts of the Highlands of Scotland to this day.

In Britain, till within these twelve years, the flail may be said to have been the only instrument employed for threshing corn; but previous to that period, several attempts were made to construct machines for performing that laborious work. The first attempt which we know of with certainty, was made by an ingenious gentleman of the county of East-Lothian, Mr Michael Menzies, who invented a machine that was to go by water, upon the principle of driving a number of flails by a water-wheel, but from the force with which they wrought, it was found the flails were soon broken to pieces, and consequently the invention did not succeed.

Another threshing machine was invented about 1758, by Mr Michael Stirling, a farmer in the parish of Dumblain, Perthshire. This machine was nearly the same as the common mill for dressing flax, being a vertical shaft with four cross arms, inclosed in a cylindrical case, three feet and a half high, and eight feet diameter. Within this case the shaft, with its arms, were turned with considerable velocity by a water-wheel, and the sheaves of corn being let down gradually, through an opening for the purpose, on the top of the box, the grain was beat off by the arms, and pressed with the straw through an opening in the floor, from which it was separated by riddles shaken by the mill, and then cleaned by fanners also turned by it. The great defect of this machine was, that it broke off the ears of barley or wheat, instead of beating out the grain, and was only fit for oats.

A third ſpecies of a threſhing mill was attempted by two gentlemen in Northumberland about 1772, viz. Mr Elderton near Alnwick, and Mr Smart at Wark, nearly about the ſame time. The operation was performed in their machine by rubbing inſtead of beating. The unthreſhed corn was carried round between an indented drum, of about ſix feet diameter, and a number of indented rollers arranged around the circumference of the drum, and preſſed towards it by ſprings, ſo that when the drum revolved, the grain was rubbed out in paſſing between it and the rollers. This machine was found, on trial, even more defective than the former, as it not only bruiſed the grain, but did very little execution, though the Northumberland ſurveyors, either from inadvertency or miſtake, would arrogate to that county the invention of the threſhing machine now in uſe, from which this attempt was obviouſly different*.

The late Sir Francis Kinloch of Gilmerton, Bart. having ſeen the Northumberland machine, attempted to improve it by incloſing the drum in a fluted cover, and inſtead of making the drum itſelf fluted, he fixed on the outſide of it four fluted pieces of wood, capable of being raiſed a little above the circumference of the drum by means of ſprings underneath, ſo as to preſs againſt the fluted cover, and rub out the grain as the ſheaves paſſed round between them; but, finding that it bruiſed the grain in the ſame manner as the Northumberland machine did, he ſent it to Mr Andrew Meikle at Know-

* In a correſpondence with thoſe gentlemen on this ſubject, they authoriſe us to ſay, that, from recent information, they are now convinced the ſtatement given by them in the Northumberland Survey is defective, *and that they are ſatisfied the merit of perfecting the machine, as ſpecified in the patent, belongs ſolely to Mr Meikle.* R. B.

mill, in his neighbourhood, in order to have it rectified, if possible.

Mr Meikle who, for several years, had been making many trials of different machines for the same purpose, after repeated experiments with Sir Francis' mill, found that it was constructed upon wrong principles, and that beating must be had recourse to, instead of rubbing. He therefore, in 1785, made a working model, turned by water at Knowmill, in which the grain was beat out by the drum, after passing through two plain rollers, which were afterwards altered for two fluted ones. Mr George Meikle, son of the former, being at Kilbegie, the residence of Mr Stein, agreed to erect a machine of this nature for that gentleman, upon condition of Mr Stein furnishing all the materials, and paying him for the work, *only in case the machine answered the desired purpose*. This was agreed to, and the machine was completed in February 1786, being the first ever made. It was found to work exceedingly well, and the only alteration made from the above mentioned model was, that, instead of plain rollers, fluted ones were substituted. In consequence of this successful attempt, a patent for the invention was applied for, which, after a considerable opposition from a person no ways concerned in the invention, was obtained in April 1788.

These machines have now spread over all the corn counties of Scotland, and have lately been successfully introduced into the northern counties of England, though, strange to tell, they are scarcely known in the southern and best cultivated parts! During our progress in the West Riding, we saw a few of them, which were wrought by 2 horses, and seemed, so far as that strength would allow, to perform the work in a sufficient way.

Where farms are of small size, it would be superfluous to recommend the erection of large machines, as they intereſt of the original outlay would be a heavy drawback from the advantages; but, under contrary circumſtances, we are decidedly of opinion, that a machine of great powers, provided with two rakes or ſhakers, and two pair of fanners, is the moſt profitable one for the poſſeſſor. By a machine of this kind, when wrought by horſes, the grain is completely threſhed and cleaned, at little more expence than is paid for cleaning it alone when threſhed by the flail, independent of the additional quantity of corn produced by the powers of the machine; and when wind or water is ſubſtituted inſtead of horſes, the ſaving is conſiderably encreaſed.

A horſe machine of the greateſt powers, with the appendages of rakes and fanners, may be erected for one hundred pounds, and when wrought by wind, for two hundred pounds independent of the buildings and fixtures which are required. It would be unfair, however, to charge theſe to the account of the threſhing machine, as, even upon a middle ſized farm, a greater extent of buildings is required for barn work, when the corn is ſeparated from the ſtraw by the flail, than when the operation is performed by the threſhing machine.

From the moſt minute attention we could beſtow on this ſubject, we are confident an extra quantity of corn, equal, in ordinary years, to five per cent. will be given by the threſhing machine more than by the flail, beſides innumerable other advantages which accompany that machine. Indeed, the loſs by the flail has long been proverbial, and the beſt of farmers were obliged to ſubmit to loſſes of this nature, becauſe they could not be remedied; but with the threſhing machine no corn need be loſt, as every particle of grain is ſcutched off, when the machine is conſtructed upon right principles.

The expence of horſe labour, from the encreaſed value of the animal, and the charge of his keeping, being an object of great importance, we beg leave to recommend that, upon all ſizeable farms, that is to ſay where two hundred acres, or upwards, of corn are ſown, the machine ſhould be wrought by wind, unleſs where local circumſtances afford the conveniency of water, which is always to be preferred. Many perſons recommend what they don't practiſe; but the ſurveyors of the Weſt Riding are not in this predicament: Upon their farms the machines are all driven by wind, and upon two of them horſe machines are annexed, which prevents every inconvenience that might ariſe during a tract of calm weather.

Wind machines were, till lately, expoſed to dangerous accidents, as the ſails could not be ſhifted when a briſk gale aroſe, which is often the caſe in this variable climate. Theſe diſagreeable circumſtances are now effectually prevented by the inventive genius of Mr Meikle, and the machine may be managed by any perſon of the ſmalleſt diſcernment or attention.

The whole ſails can be taken in, or let out in half a minute, as the wind requires, by a perſon pulling a rope within the houſe, ſo that an uniform motion is preſerved to the machine, and the danger from ſudden ſqualls prevented.

Where coals are plenty and cheap, ſteam may be advantageouſly uſed for working the machine. A reſpectable farmer in the county of Eaſt Lothian works his machine in this way, and being ſituated in the neighbourhood of a coalliery, is enabled to threſh his grain at a trifling expence.

The quantity of grain threſhed in a given time, muſt depend upon its quality, on the length of the ſtraw, and upon the number of the horſes, or ſtrength of the wind, by which the machine is wrought; but under favourable circumſtances, from eighty or ninety buſhels of oats, or from forty or ſixty buſhels of wheat, may be threſhed and cleaned in one hour. This we can speak of with certainty, because we have threshed the above quanities ourselves; but it is from clean dry grain only that so much will be done in that period.

In a word, the threshing machine is of the greatest utility to the farmer, and from it the public derives a vast additional quantity of food for man and beast. If *five* per cent. is added to the national produce, it is as great a gain as if the national territories were increased one seventh more than their present size, for this additional quantity of grain is produced without any other expence than the money laid out in erecting the machines, no more seed is sown than formerly, nor more labour employed, and these articles, with the rent, have always been taken as equal to two thirds of the produce.

If these things be true, and we are confident whoever is acquainted with the subject, and seriously investigates the extent of the beneficial consequences arising from the threshing machine, will acknowledge them as facts, we beg leave to say too much cannot be done towards rewarding the inventor. Mr Meikle has hitherto, from causes unnecessary to mention, received little benefit from his patent right, which has been scandalously encroached upon; and if public munificence should ever be employed in rewarding the authors of useful inventions, it cannot be bestowed upon a person who has a greater claim.

As a farmer's capital ought never to be laid out in expensive building, or works of an extraordinary kind, we are humbly of opinion, that the sums necessary for erecting machines, should, in the first instance, be expended by the landlord, and the tenant taken bound to leave them in a *workable condition* at his departure. Many farmers have capitals sufficient for undertakings of this kind, but the great body of that profession would be injured by such outlays, as they would thereby be deprived of the means of improving their farms in other respects. Besides, as every improvement, at the long run, centers in the pocket of the proprietor, it is but fair and reasonable he should contribute his moiety of the expence laid out in procuring it; and in many cases he would be benefited in the first instance by the erection of threshing machines, particularly where new farm steadings are to be built, as fewer houses would of course be necessary.

Mr Meikle's patent right having been lately called in question we beg leave to say a few words more in his favour. If any machine constructed upon similar principles to those contained in the specification of his patent was previously erected, let it be pointed out, and we will give up the cause. The old Northumberland machine, which did not thresh, but bruise out the corn, is now laid aside, while Mr Meikle's, or, which is the same thing, the works of those who have stolen his invention, have circulated over more than one half of the island. The Northumberland surveyors say, that the leading principle of the invention was taken from the flax mill, and mention, that one Mr Gregson had used a machine, constructed in imitation of it, for threshing his grain. According to their own account, this machine was useless, as it did not thresh so much in a day as a good barn man could do in the same period; and it is evident from being soon laid aside, that such a machine was incapable of executing the arduous task of threshing with advantage. Allowing even, for argument's take, that the first idea of the threshing machine was taken from the flax mill, it proves nothing against Mr Meikle's right; for every invention whatever is drawn from some source or other. Mr Meikle was the first man that constructed a machine capable of separating the grain from the straw in a proper maner. We speak with confidence on this subject, because we have had the best opportunities of knowing it in every ſtage, and are warranted in attributing the merits of the invention ſolely to Mr Meikle, and of declaring that the Britiſh nation, and the whole world, are indebted to him for a machine which enſures the moſt laſting and important advantages to the intereſts of agriculture.

The following Letter from an extenſive farmer in this county, addreſſed to Sir John Sinclair, Bart. corroborates what we have ſaid reſpecting the utility of the threſhing machine, and the merits of the inventor.

" Agriculture is the antienteſt, as well as the moſt valuable of the ſciences, and will always be conſidered, by every wiſe government, as an object of primary attention. In Britain, the cultivators of the ground have had too much cauſe to complain, that while trade and manufactures experienced the foſtering hand of the legiſlature, the internal improvement of the country was neglected and undervalued. The eſtabliſhing a National Board of Agriculture has, however, in part done away this complaint; and it remains with you, and the other reſpectable Members of that Board, to render the inſtitution ſalutary and uſeful.

" Fully impreſſed with a ſenſe of the beneficial conſequences which may accrue from a National Board of Agriculture, conducted upon proper principles, I beg leave to call your attention to the leading objects of ſuch an inſtitution. Theſe, in my humble opinion, conſiſt in uſing every endeavour to remove obſtacles to improvement which exceed individual ſtrength, and in rewarding and encouraging the authors of uſeful inventions, whereby the practical or operative department of Agriculture is facilitated or improved.

" The firſt of theſe objects I do not mean at this time to enter upon; but, with reſpect to the ſecond, a doubt cannot be entertained as to the propriety of its occupying a principal ſhare of your deliberations. I am well aware, that the paucity of your funds effectually prevents you from beſtowing premiums or rewards in the firſt inſtance; but this does not hinder the Members of the Board from diſcharging their duty, by recommending the caſe of meritorious perſons to the conſideration of thoſe who more immediately hold the ſtrings of the national purſe. He who benefits the public, is entitled, on every principle of policy and juſtice, to a public reward; and by whom can his merits be more juſtly eſtimated than by the Members of a Board eſtabliſhed for the expreſs purpoſe of ſuperintending and promoting improvements in that very ſcience which he has benefited? In this point of view, I ſubmit to your conſideration an important improvement in a chief branch of rural economy, made by a humble but worthy individual.

"The trouble and loſs attending the ſeparation of the corn from the ſtraw, according to the old way of doing it by the flail, are ſo well known that it would be ſuperfluous to deſcribe them. This operation is now completely performed by a machine, which, in a great meaſure was invented, and, without diſpute, was brought to its preſent ſtate of perfection by Mr Andrew Meikle, engineer at Houſton mill, near Haddington, whoſe family ſeems to poſſeſs a kind of hereditary right to genius and invention, and whoſe father firſt introduced the barley mill and fanners into Scotland, in the year 1710, under the patronage of that illuſtrious character, Andrew Fletcher, Eſq; of Salton.

"If theſe machines were not ſo well known, I would enter upon a detail of their principles and powers; but, preſuming the Board are not unacquainted with theſe things, I ſhall confine myſelf to Mr Meikle's claim for receiving a national reward; and this I ſhall demonſtrate, by ſhowing the great ſavings ariſing from the invention, and the conſequent increaſe of agricultural produce.

"The firſt threſhing machine erected by Mr Meikle was completed in the year 1788; and ſince that time he has progreſſively introduced a variety of improvements, all tending to ſimplify the labour, and to augment the quantity of work thereby performed. When firſt erected, although the corn was equally well ſeparated from the ſtraw, yet, as the whole of the ſtraw, chaff, and corn, were indiſcriminately thrown into a confuſed heap, the work could only with propriety be conſidered as half executed. By the addition of rakes or ſhakers, and two pairs of fanners, all drove by the ſame machinery, the different proceſſes of threſhing, ſhaking, and winnowing, are now all at once performed, and the corn immediately prepared for the public market. When I add, that the quantity of corn gained from the ſuperior powers of the machine is fully equal to a twentieth part of the crop, and that, in ſome caſes, the expence of threſhing and cleanſing the corn is conſiderably leſs than what was formerly paid for cleaning it alone, the immenſe ſavings ariſing the invention will at once be diſcerned.

"I ſhall now offer ſome calculations relative to the probable amount of the ſavings which might accrue to the public, if threſhing machines were univerſally uſed. I do not affect to be accurate in theſe calculations, which cannot be expected before the facts are ſufficiently aſcertained; but, to borrow the words very properly uſed by you in your ſpeech to the Board, July 29, 1794, "to be enabled to form ſome general idea of the nature and extent of public improvement, is a great ſtep gained."

"The extent of ground annually employed in Britain, in the raiſing of corn, may be computed at ſeven millions five hundred thouſand acres, and the average produce of the different grains at three quarters per acre, as, below that increase, no farmer can raise it with profit. I observe, in your speech to Parliament, when you moved the establishment of the Board, that you supposed there were only five millions of acres annually employed in raising grain: but I have reason to think this is a mistake; for, if the population of the island be eight millions, the produce of these acres would be far below what is required for the support of that number of people, independent of what is necessary for the feeding of horses and sowing the next crop. I observe also, in the reprinted survey of the county of Stafford, a pretty just calculation of the number of acres annually sown in that county, amounting to one hundred and fifty thousand acres. Now, as Stafford is not a corn county, I do not take much latitude when I fix upon it to average the whole counties of England; this would make the total quantity sown in that kingdom amount to six millions of acres. The remaining one million five hundred thousand acres I suppose to be sown in Scotland and Wales, which makes their produce only equal to that of ten English counties.

If seven millions five hundred thousand acres be annually sown in Britain, and the average produce amount to three quarters per acre, then the total quantity of grain annually raised in Britain would be twenty-two millions five hundred thousand quarters.

I have already said, that the threshing machine, from its superior powers, will give one twentieth more grain than when the operation of threshing is performed by the flail, which, from any trials I have made, will be rather exceeded: this gives an increased quantity of one million one hundred and twelve thousand five hundred quarters; which, taken at the average price of thirty-

two-shillings per quarter for all grains, amounts to - - - - - -	L. 1,781,250
Add to this the difference of expence between threshing with the above machine and the flail, which may be stated at 1s. per quarter, although, when the machines are wrought by wind or water, the difference is more than double that sum. This, on 22,500,000 quarters, is	1,125,000
	L. 2,906,250

"I scarce expect to be credited when I say, that the above enormous sum would annually be saved to the public, if the whole corn annually raised in Britain was separated from the straw by these machines, and yet few political calculations will admit of such certain demonstration. Let me only suppose, that one eighth of our corn is threshed in that way, and still the saving is immense. If any person doubts the principles upon which these calculations are built, I have only to request he would pay strict attention to the subject, and I am pretty positive he will soon acknowledge they are not overstretched. The only deduction necessary to be made, is for the interest of the money expended in erecting the machines; the principal sum of which, especially upon large farms, will be repaid by the savings of three years crops.

"If it be the object of a National Board of Agriculture to reward and encourage the authors of useful inventions in the operative department of that science, (as I think it is) where is the man who deserves a greater share of their favour than the ingenious mechanic I have mentioned? Mr Elkington, at their recommendation, received a Parliamentary grant of one thousand pounds, and probably he deferved it; but without meaning to derogate from the merits of that gentleman, I will not affront Mr Meikle fo much as to put the invention and improvement of the Threfhing Machine into the fcale with the new mode of drainage.

" Perhaps a ftranger, upon reading this letter, may exclaim, " What! has the author of this ufeful invention received no reward? Has the man who leffened the toil of human labour, who devifed the means of encreafing the ftock of agricultural produce, and confequently augmented the national wealth, received no mark of public favour?" No he has not; unlefs a patent-right of fourteen years to erect thefe machines, the greateft part of which is expired, can be confidered as fuch;—I may add, that owing to certain circumftances, Mr Meikle has hitherto received little or no benefit from the patent; and if the fees of office be taken into account, I am juftified in faying, he had better have remained without fuch a right.

" That every increafe of agricultural produce, and every faving of the expence of farm labour, ultimately centre in the pockets of the landed proprietor, I confider as an incontrovertible propofition. Now here is a great increafe of produce, and an immenfe faving of labour, all flowing from the unabated efforts of an individual, whofe intereft, confidering the limited circle in which he moves, can fcarcely be benefited from the invention, unlefs he participates of legiflative munificence. If any perfon were to devife a fcheme, from which, the monied intereft of the kingdom could legally reap double intereft, upon their bonds, bills, &c. what obligations would that clafs of the community confider themfelves to be under to the author of fuch a fcheme? and yet the landed intereft of Britain receive greater advantages from the invention of the threſhing machine, and, ſtrange to tell! have totally neglected the merits of its worthy inventor.

"May I therefore hope, Sir John, that you, and the other reſpectable Members of the Board of Agriculture, will take this buſineſs under your conſideration. By procuring a reward for Mr Meikle, you will not only diſcharge a debt incumbent upon the whole landed intereſt of the kingdom, but will alſo ſtimulate other ingenious mechanics to uſe their utmoſt endeavours to ſimilar improvements."

NOTES on Chap. 3.

(*a*) I am rather inclined to think, that the unſteadineſs of theſe ploughs is owing to their *bad conſtruction* in *other reſpects*, as I have ſeen ploughs with the beams placed in this manner, going well, where the horſes were yoked *one before another*; and it is capable of demonſtration, that it is (to a certain degree) a proper poſition for the beam *when the horſes are ſo yoked*; but it is certainly a very improper one, when the horſes are yoked *double* or abreaſt, which they always ought to be, except in ſome particular caſes.

Mr Bailey.

(*b*) I believe if we could get into the *habit* of one-horſe carts, it would be an advantage, but we have got the habit of three-horſe carts rivetted on us. *W. P.*

CHAPTER VI.

INCLOSING.

THE whole of the West Riding is inclosed except the common fields and moors, and too much praise cannot be bestowed on the perfect state, in which the fences are kept. The inclosures are, however, generally too small, at least for corn fields, and at any rate occasion a great waste of ground. It did not appear to us, that either the conveniency of water, or uniformity of soil, had been much studied in laying them out (*a*); these are objects of importance, and without paying suitable attention to them, the full advantages of inclosing cannot be attained.

We bestowed great pains in endeavouring to ascertain, how much the rent or value of the ground was increased by a regular inclosure, and from the information we received, it amounted at least to 25 per cent. Many speculative men have asserted, that the inclosing of ground is injurious to the public (*b*); that it tends to depopulate the country; that it serves to render corn scarce and dear, and is prejudicial to the lower ranks. We shall say a few words on each of these points:

1st, That inclosing of ground, cannot be injurious to the public, is evident; as it occasions an immediate rise of rent to the landlord; and how could the raised rent be paid, if more corn and grass were not produced by this change of system, than under open field management? Inclosing enables the farmer to practice every improvement; it gives him an opportunity to introduce the grafs hufbandry in all its perfection, and to depafture his fields, with fuch kind of flock as they are naturally adapted to. In a word, without inclofures, a farmer can fcarcely manage his poffeffion in an advantageous way, or cultivate it in a manner fuitable to its different qualities and fituations.

If thefe things are true, the public good muft neceffarily be promoted by every judicious inclofure which is made. What is the public good but the good of individuals accumulated? Inclofing raifes the rent payable to the landlord; is favourable to the intereft of the tenant, and enables him to carry on his bufinefs with judgment and accuracy: It increafes the food of the people, as more corn and grafs are produced under this mode of management, than under that of open field; and gives employment to many perfons, who would otherwife have remained idle and ufelefs members of the ftate.

2dly. It is faid, inclofing tends to depopulate the country. During our furvey, *we repeatedly made enquiries* upon this point, and were uniformly *anfwered*, that inclofing increafed population. This is fo contrary to the opinions of fome popular writers, particularly the late Dr Price, that it cannot be improper to inveftigate the bufinefs.

Thefe gentlemen argue upon the fuppofition, that the moment a field is inclofed, it muft neceffarily be kept in grafs, which they fagacioufly think, gives employment to few hands, or, as they commonly exprefs it, only to a fhepherd, and his dog. They do not reflect, that the fame quantity of land, if not more, would be kept in grafs, whether there was a fingle inclofure or not; as cattle and fheep muft be fed, one way or other, equal to the demand. Whenever more land is in grafs than the demand for thefe articles requires, the fyftem muft immediately be changed, as the prices of butcher meat, are so high in Britain, that an exportation of it can seldom take place. This is not the case with corn, for the bounty given on exportation, enables our merchants to send it abroad, when a superfluity remains at home.

By inclosing of land, the quantity necessary for producing as much grass as will feed cattle and sheep for supplying the market is reduced. We are inclined to think, this position will not be questioned by any person who considers the rapidity with which beasts feed in a proper inclosure, in comparison to those herded in an open field. This consequently leaves more land to be cultivated for corn, and, upon their own principles, inclosing must prove friendly to population.

Another thing which has escaped the notice of these gentlemen, is the number of people who receive employment from the hides and skins of the animals depastured on grass land. While they examine the field they perhaps don't see a single person amongst the bestial: Hence they set down at once, that the grass system is destructive to the population of the country. But let them consider the number of curriers, shoemakers wool-combers, and manufacturers, who are thereby provided in work, and they will allow, that an acre of grass affords employment to as many people as an acre of corn land. This point is so clearly elucidated in the Hereford Survey, that we beg leave to refer the candid inquirer to it for a full proof.

3dly, *Inclosing serves to render corn scarce and dear:* If what we have already mentioned be just and well founded, the reverse is the consequence of inclosing. When land is, for a few years, refreshed with grass the crops of corn which it then produces, will nearly be doubled. This fact is so well known that it would be superfluous to support it by arguments.

4thly, Inclosing is prejudicial to the interest of the lower ranks. This, if it has any meaning at all, can only happen where waste land is inclosed, on the margins of which incroachments have been made; many cottagers in these situations keep half-starved cows, geese, &c, in the herding and attending of which, they consume more time than the small advantage they receive can compensate. But if their possession gives them a legal right of servitude, they are entitled and will receive their share when a division takes place. If they have gone beyond their right, and eaten with their beasts what was the property of their neighbour, this affords no reason why the encroachment should be perpetual.

Upon the whole, we are clearly of opinion, that inclosing of land is of great public advantage, that it cannot decrease population, but, on the contrary, by furnishing food and employment, must materially contribute to increase the number of the people; that it is the means of rendering corn plentiful, and cannot be prejudicial to the lower ranks. These things will likely be unanimously acknowledged by all practical men who take the trouble to examine the common fields, and those numerous and immense tracts of waste ground, which, to the shame of this country, remain comparatively unproductive to the state (c).

Respecting the size of inclosures, it would be improper to lay down any particular rules, as this should be regulated by the size of the farm on which they are made. In general, it may be remarked, that where a regular rotation of cropping is followed out, they should be of such a size as may be sown in one year with the same grain, or that the strength kept on the farm can fallow in one season. Also, we remark, that the larger the inclosure, the cheaper it is executed, and the less ground left unproductive.

As to the manner of inclosing, we know of no fence equal to a good quick-set hedge of white thorn, when it is properly trained up. Thorns when planted on a clean soil, and fenced with post and rail for a few years will soon produce a complete hedge. Perhaps stone walls are more eligible where sheep are kept. These we would recommend to be built, or rather lipped with lime, and to be six quarters in height, with an additional quarter by way of caping. Probably this, at the long run, is the cheapest fence; but, being very expensive at first, it should in every case be executed by the proprietor, the tenant paying legal interest upon the outlay's.

NOTES on Chap. 6.

(*a*) This is a moſt eſſential point. In my farm I have not one field but what is of two or three natures, yet all muſt be under one mode of management; though it is naturally impoſſible it ſhould be all fit at one time. One part being a fine ſandy ſoil, fit for turnips, which may be wrought at any time; another, wet and ſtrong, and only at particular times proper for horſes to come upon. This ſhould be duly conſidered in making new incloſures, where lands of the ſame quality ſhould be laid together, not only for the convenience of the farmer, but as it is a conſiderable advantage to agriculture, and of much more importance, than uniformity or regularity of incloſures.

A Yorkſhire Farmer.

(*b*) Want of incloſing is ever the cauſe of declining population. Want of rural work drives young people from the ſalubrious and invigorating labours of the fields, to the peſtilent and deſtructive air of manufactories and great towns. It is evident from all the bills of mortality of the great towns, that they would be deſerts in 20 years, without conſtant ſupplies of young people of both ſexes from the country.

W. P.

(*c*) "Incloſing" (ſays the great Linnæus) "is the only means of having any valuable improvements carried on effectively; but our landlords and farmers are equally averſe to any expences beyond thoſe certain ones of the day, which they cannot eſcape; now this can only be remedied by the legiſlative power, which ought to oblige all proprietors to incloſe their fields in ſome ſubſtantial manner;" (and the preſent waſtes alſo) "and to enable them, at the ſame time, to raiſe their rents upon their tenants, ſufficiently to pay good intereſt for the ſums expended."

CHAPTER VII.

ARABLE LAND.

BEFORE entering upon this chapter, we think it necessary to make some preliminary observations, so as the different systems practised in this extensive district may be easier understood.

1st, A great part of the West Riding is exclusively kept in grass, and where this is the case, cultivation by the plough is considered as a secondary object.

From Ripley, to the western extremity of the Riding, nearly the whole of the good land is kept under the grazing system, and seldom or never ploughed, while corn is raised upon the inferior or moorish soils. During the time we were in that part of the country, we hardly ever saw a plough; and a stack of corn was a great rarity. Upon the higher grounds, there are immense tracts of waste, which are generally common among the contiguous possessors, and pastured by them with cattle and sheep. Some of them are stinted pastures, but the greatest part are under no limitations: the consequences of which are, the grounds are oppressed, the stock upon them starved, and little benefit derived from them by the proprietors.

2dly, The land in the vicinity of manufacturing towns. The greatest part of the ground is there occupied by persons who do not consider farming as a business, but regard it only as a matter of convenience. The manufacturer has his inclosure, wherein he keeps milch cows for supporting his family, and horses for carrying his goods to market, and bringing back raw materials. This will apply to the most part of the land adjoining to the manufacturing towns; and although much ground is not, in this case, kept under the plough, yet comparatively more corn is raised, than in the division above described.

3dly, The corn district, or those parts of the Riding where tillage is principally attended to, and grass only considered as the mean of bringing the corn husbandry to perfection.

If we run an imaginary line from Ripley southward by Leeds, Wakefield, and Barnsley, to Rotherham, we may affirm, that the greatest part eastward of it, till we come to the banks of the Ouse, which separates the West from the East Riding, is principally employed in raising corn. About Boroughbridge, Wetherby, Selby, &c. there is about one half of the fields under the plough. Further south, about Pontefract, Barnsley, and Rotherham, there are two-thirds; and to the eastward of Doncaster, to Thorn and Snaith, three-fourths of the land are managed in a similar way. There is not much waste in this division, but what is in that situation, is capable of great improvement.

4thly, The common fields. These are scattered over the whole of the last division, but are most numerous in that part of the country to the eastward of the great north road, from Doncaster to Boroughbridge. It is impossible even to guess at the quantity of land under this management. In general, it may be said to be extensive, and from the natural good quality of the soil, and the present imperfect state of culture, great room is afforded for solid and substantial improvement being effected upon all land coming under the description of common field (*a*).

5thly, The moors. These, besides the large tracts in

the first division, lie in the western part of the Riding, and perhaps contain one-eighth of the district. Upon them sheep are chiefly bred, and afterwards sold to the graziers in the lower parts of the country. A great part of them is common, which lays the proprietors under the same inconveniences as are already pointed out; and which might easily be remedied, by dividing and ascertaining the proportion which belongs to the respective proprietors (*b*).

Having given these preliminary observations, which we trust will afford a general idea of the present state of husbandry in this district, we shall now proceed to detail the different articles included in this chapter.

SECT. I.—*Tillage.*

THE West Riding cannot be considered as a district where the cultivation of corn is practised in the most approved way, and many circumstances concur to retard its improvement. From the flourishing state of manufactures, capitals are thrown into that line, which in other places would be employed in the cultivation of the soil; and the advantageous markets for disposing of cattle and sheep, induces the actual farmer to bestow a greater portion of his attention upon the management of his live stock, than upon his corn fields. This observation we make in justice to the farmers of the West Riding, many of whom have their farms in the most perfect condition. Where the case is different, it is but fair to infer, that the above mentioned circumstances have operated to prevent them from being so perfect as their neighbours.

The arable soils of this Riding, as referring to cultivation, may be considered as comprehending all the varieties which prevail in Britain, but the prevailing quality (keeping off the moors) is loam, the value of which is in a great measure regulated by the subsoil, upon which it is incumbent; limestone land, or in other words, where the surface lies upon a limestone bottom, is also very prevalent, and a great part of that large tract of ground adjoining to the river Ouse, is of a clayey tenaceous nature, holding water like a cup, very difficult to manage, but, under the hands of skilful cultivators, capable of carrying the most luxuriant crops.

Every kind of grain, pulse, roots, and other vegetables, cultivated in the fields, are produced in the West Riding but a particular, account of these shall be given in the fourth section of this chapter.

SECT. 2.—*Fallowing Defended.*

WHETHER summer fallow is necessary or unnecessary? is a question lately agitated; and in a respectable work, (the Survey of Norfolk) an attempt has been made to explode this practice, which has long been considered as a most beneficial improvement. The agriculture of Britain being materially interested in the issue of this question, the following answers to the Norfolk surveyor, are submitted to the public.

To keep his land clean will always be a principal object with every good farmer; for, if this be neglected, in place of carrying rich crops of corn or grass, the ground will be exhausted by crops of weeds. Where land is foul, every operation of husbandry must be proportionally non-effective, and even the manures applied, will in a great measure be lost.

If the season of the year, and the state of the weather, when the ground is ploughed, preparatory to receiving the seed, be duly considered, it will be found, that at that time, it can neither be properly divided by the action of the plough; nor can root weeds, or annual weeds, be then extirpated. Hence arises the necessity of working it in summer, when the weather is favourable for the purposes of ploughing, and when root weeds may be dragged to the surface. It is only at that time the full advantages of ploughing are attainable; for summer fallow may with propriety be stiled ploughing in perfection.

The necessity of summer fallow, depends greatly upon the nature and quality of the soil, as upon some soils a repetition of this practice is seldomer required than upon others. Wherever the soil is incumbent upon clay, or till, it is more disposed to get foul, than when incumbent upon a dry gravelly bottom; besides wet soils, from being ploughed in winter, contract a stiffness which lessens the pasture of artificial plants, and prevents them from receiving sufficient nourishment. When land of a dry gravelly quality gets foul, it may easily be cleaned without a plain summer fallow; as crops, such as turnips &c. may be substituted in its place, which, when drilled at proper intervals, admit of being ploughed as often as necessary; whereas wet soils, which are naturally unfit for carrying such crops, must be cleaned and brought into good order by frequent ploughings and harrowings during the summer months.

It is from neglecting to make these distinctions, that the erroneous system laid down by Mr Kent, the Norfolk Surveyor, evidently proceeds.

The county of Norfolk generally consists of dry sand, or of rich sandy loam; and, agreeably to the above principles, summer fallow may in that district be considered as unnecessary. If Mr Kent had confined his strictures to the husbandry of Norfolk, no objection could reasonably have been urged against them, but when he condemns summer fallow altogether, he strikes at the agriculture of Britain in a most material point.

The substance of Mr Kent's arguments against fallow, may be comprised under four heads:

1st, Nature does not require any pause or rest, and the earth was evidently designed to yield a regular uninterrupted produce.

2dly, As the productive quality of the earth never ceases, if corn is not sown, weeds will be produced; therefore it is our business to expel the unproductive plant, and to introduce others that are beneficial.

3dly, That the idea of leaving land to rest is ridiculous, for by keeping it clean, and by a judicious intermixture of crops, it may be managed like a garden, and sown from one generation to another.

4thly, That the fallows in England exhibit nothing but a conflict betwixt the farmer and his weeds, in which the latter generally prevail, for they are only half stifled, and never effectually killed.

The most of these arguments may be granted, and yet the utility, nay, the necessity of summer fallow be consistently maintained.

It is already acknowledged, that it is only upon wet soils, or in other words, upon land unfit for the turnip husbandry, a plain summer fallow is necessary, and this we suppose includes three fourths of the island. The utility of summer fallow upon such soils is not contended for because nature requires a pause or rest, to invigorate her to carry fresh crops, but solely because it is impossible to keep them clean without this auxiliary assistance. To speak of following nature in farming is mere sound; for if we were to imitate nature, we would not cultivate land at all. Nature is often improved by art, and fallowing is the means employed for removing a host of enemies, which prevent her from being fertile and productive.

As a field filled with root weeds, must be in a state of greater exhaustion, than if it carried a heavy crop of corn, so the productive quality of the earth must necessarily decrease in proportion to the quantity of weeds it brings forth. But because corn is not sown, it does not follow that weeds of any kind should be suffered to grow. The object of allowing the ground to remain a year under fallow, is to afford time and opportunity for expelling the unproductive plant, and to prepare it for the reception of others, which are beneficial.

The most judicious intermixture of crops upon clay soils, will not preclude the necessity of summer fallow, although it will go a great way to prevent a frequent repetition of it. An eighth course shift, such as fallow, wheat, beans drilled and horse-hoed, barley, grass seeds, oats, beans, and wheat, is as much as can be recommended, and it is only upon rich clay, or deep loam, where such an extensive rotation is admissible. A shift of this kind, when dung is applied twice in the course of it, will pay the farmer more handsomely than the most judicious intermixture of crops, where fallowing is neglected.

Again, no rules drawn from garden practice, will apply to operations carried on in the field; the soils are generally very different, and any comparison that can be made, must be with these rich sandy soils, upon which we have allowed fallowing to be unnecessary. The crops in the garden are reaped at so many different times, and often so early in the season, that opportunity is always gained for working the ground in the compleatest manner, while the immense difference betwixt working with the plough and the spade renders every comparison ridiculous.

A fallow field which exhibits a conflict betwixt the farmer and his weeds, does not deserve that appellation; for the intention of the fallow is to extirpate these weeds. We are inclined to think, that the shocking situation of many English fallows may be attributed to the feeding, and folding them with sheep. The farmer, from being obliged by the conditions of his lease, or the rules of common field management, to fallow every third or fourth year, is tempted to draw something from them when in this unproductive state, and, to gratify his avarice in the first instance, sacrifices the good husbandry which it is his ultimate interest to practice. A well managed fallow should be wrought as early in the season as possible, and continually turned over where the least particle of quickens appears. It is no argument against the utility of fallows, that they are often managed in a different way; this goes only against the impropriety of the management, but does not militate against the practice itself.

Upon the whole, the necessity of summer fallow turns upon this single point. Can wet lands be advantageously employed in raising turnips or cabbages? a question which the *practical farmer*, who is sufficiently acquainted with the nature of such soils, and the immense labour required to bring them into proper tilth, will have no difficulty to answer in the negative. It is not disputed but that turnips and cabbages will grow upon these soils; but the question is, whether the extraordinary labour they require, and the damage sustained by the ground, during the consumption or carrying off the crops, will not exceed the value of the produce? Does Mr Kent mean to recommend the turnip husbandry under such circumſtances? If he does, the recommendation furniſhes a preſumption that, he is unacquainted with the cultivation of wet lands. If he does not, how is the ground to be kept clean, and enabled to yield a regular uninterrupted produce?

Nothing that is ſaid in defence of Fallow, is meant in vindication of the abſurd ſyſtem of taking only two crops to one fallow, as practiſed upon many Engliſh common fields. It is only meant to ſhow that clay ſoils, and every ſoil incumbent upon a wet bottom, cannot be kept clean, without the aſſiſtance of this radical and antient practice. How often it ſhould be uſed, muſt in a great meaſure be left to the diſcretion of the farmer, who will repeat it when neceſſary if he knows his own intereſt. We ſhall conclude our defence of fallow, with an extract taken from p. 192 of the Survey alluded to. '*It is highly proper to be careful againſt adopting the viſionary recommendations of modern theoriſts, who, upon hypotheſes of their own, bold up wild ſyſtems of deluſion, which are apt to miſlead the credulous, and do great injury.*'

As many different opinions prevail relative to the manner in which a Fallow ſhould be conducted, we beg leave to ſtate our ſentiments upon that head.

Upon all clay ſoils (and upon ſuch only, we underſtand a complete ſummer fallow to be neceſſary) the firſt ploughing ought to be given during the winter months, or as early in the ſpring as poſſible, which promotes the rotting of the ſward and ſtubble. This ſhould be done by gathering up the ridge, which both lays the ground dry, and rips up the furrows. As ſoon as ſeed time is over, the ridge ſhould be cloven down, preparatory to to croſs ploughing; and, after lying a proper time, ſhould be harrowed and rolled repeatedly, and every particle of quickens that the harrows have brought above ſhould be carefully picked off with the hand. It is then proper to ridge or gather it up immediately, which both lays the land in proper condition for meeting bad weather, and opens up any faſt land that may have been miſſed in the furrows when the croſs ploughing was given. After this harrow, roll, and gather the root weeds again; and continue ſo doing till the field is perfectly clean (e).

We obſerve that the celebrated Mr Marſhall, in his Treatiſe upon the Yorkſhire Huſbandry, recommends a practiſe quite different. In his opinion, ploughing only neceſſary, and taking out live roots by the harrow, and carrying them off, is an evident impropriety. Mr Marſhall lately uſed ſimilar arguments to one of us who had the pleaſure of a perſonal converſation with him. We ſhall therefore do our beſt endeavours to obviate his arguments.

Frequent turning over the ground, although abſolutely neceſſary while the proceſs of fallowing is going on, can never eradicate quickens, couch-graſs, or other root weeds. In all clay ſoils, the ground turns up in lumps, which the ſevereſt drought will not penetrate, or at leaſt not ſo far as to kill the plant contained in the heart of them. When the land is ploughed again, theſe lumps or clods are ſimply turned over, and no more; and the action of the plough ſerves in no ſhape to reduce them, or at leaſt in a very imperceptible manner. If ever there was a ſeaſon for making good fallow by ploughing, it was that of 1793; there was hardly a drop of rain the whole ſummer; the drought was exceſſive, and attended with an almoſt continued ſun ſhine. Notwithſtanding all theſe advantages, the fallows which were not properly reduced in the beginning of the ſeaſon, took on a growth as ſoon as moiſture came, about the beginning of harveſt. Even when they were completely harrowed and rolled, it was found difficult to extirpate couch, as the dryness of the ground did not allow it to part so well from the clod as in seasons more moist.

If this was the case in such a dry season as 1793, what would the consequences be if the fallows were at all times to be wrought with the plough, without attempting to drag the roots to the surface by the operation of harrowing? In wet weather, the land might appear black above for a few days; but the enemy, being still in the house, would soon make his appearance. By carefully gathering all the root weeds, when the land is reduced by harrowing, which on many soils is only practicable after the roller is used, an enemy is converted into a friend; for if the stuff so gathered is accumulated into a heap, frequently turned over, till it rots, and mixed with lime, a most excellent compost is produced (*d*).

There is very little danger that clay land will ever be too much reduced by the different harrowings and rollings proposed to be given; as the last furrow, if taken deep, will raise a mould sufficiently rough for covering the seed, and for protecting the wheat during the winter. Upon such soils, nothing but frost will reduce and mellow the land perfectly; and we have seen the necessity of leaving fields of this description to be wrought in the spring, from the absolute impossibility of eradicating or killing the couch, till reinforced by this powerful auxiliary.

We shall just mention another argument in favour of gathering root weeds:—that in no other way can the purpose for which fallow is intended, be so cheaply attained. Every furrow that is given, will at least stand the farmer 7s. per acre; and if hand gathering will save one single ploughing, its expence is amply repaid; while at same time we contend, that more root weeds are taken off by gathering them once, than will be destroyed by a couple of ploughings, allowing the ſeaſon to be ever ſo favourable.

We have heard of ſome other writers, that condemn clean ſummer fallow altogether, as an unneceſſary waſte of rent and labour; which, in their opinion, might be ſaved, and the ground kept in perfect good order by a proper rotation of crops. We apprehend upon all clay ſoils this is impoſſible; as every farmer who poſſeſſes ſuch ſoils, knows by experience the difficulty of keeping them clean, even with the aſſiſtance of ſummer fallows (*e*). They are ſo often ploughed wet, from neceſſity, that a ſourneſs and adheſion are contracted, which cannot be corrected without expoſing it to the hot ſummer ſun, and reducing it by frequent ploughings and harrowings. No crop can be ſubſtituted in place of fallow, for turnips are deſtruction itſelf (*f*). Drilled beans, as is already ſaid, will do well as an aſſiſtant to fallow; but however much this crop may tend to *keep* land clean, that is already in good order, we apprehend, from the neceſſity of ſowing them early, they will never anſwer as a ſubſtitute for one of the moſt radical of all improvements,—a clean ſummer fallow.

But want of fallows is not the want of the Yorkſhire huſbandry; in the corn diſtrict they prevail to a much greater extent than neceſſary, and, unleſs where turnips can be introduced, occaſion great drawback upon the farmer's profits. If good land be fallowed properly, can it ever be ſuppoſed neceſſary to repeat it after carrying only wheat and beans? When this practice is too often repeated, it alſo loſes much of its effects, the ſuperior advantages ariſing from a firſt fallow being well known to all farmers; and while we condemn the ſyſtem that would throw out this beneficial practice altogether, we are decidedly againſt an unneceſſary repetition of it.

SECT. 3.—*Rotation of Crops.*

OWING to the limitations upon management, the general rotations cannot be so liberal, or so properly adapted to good farming, as in other circumstances might be expected. Where, as fallow is required after two crops, (which is a general covenant,) no wise rotation can be introduced upon heavy lands. In that case wheat is taken after the fallow, which is succeeded either by oats or beans: where the soil answers for a fallow crop, such as turnips, barley is usually taken next, after which follows clover and wheat. This we consider to be a good rotation, where the turnips are properly cleaned; but upon loams the rotation might be much further extended, if not prohibited by covenants, as shall be afterwards explained. In the western parts of the Riding, oats are the prevailing crop, which is indeed very proper, so long as the plough is confined to the higher grounds (*g*).

As no general description of husbandry can show the particular rotation of crops in an accurate manner, we shall give a circumstantial detail of the œconomy of several farms, situated in different parts of the Riding, which will afford much practical information.

FARM, No 1. *situated in the centre of the Riding.*

Extent, 150 acres. 60 acres whereof are dry turnip soil; the remainder a mixture of clays with gravel; and incumbent on a wet bottom.

Servants, two men and a boy in the house; and two labourers for threshing, &c.

Horses	6
Milch Cows	4
Ewes -	60
Hogs -	20
Year-old Heifers	6

M

Diſtribution of crops for 1796, and the number of acres ſown, diſtinguiſhing each grain.

Wheat - -	30 acres
Barley - -	20
Oats - -	14
Meadow-graſs -	7
Red clover - -	14
Paſture - - -	45
Summer fallow and turnips	20
	150

FARM, No 2. *in the weſtern part of the diſtrict.*

Extent, 80 acres

Annual Crops,

8½ acres of oats.

½ acre of barley.

21 acres of meadow cut for hay.

20 acres paſtured with feeding cattle.

30 acres paſtured with milch cows, young cattle, and horſes.

The male ſervants kept, are one man and a boy in the houſe, and a labourer or two occaſionally; 3 horſes are kept for work, and a mare for breeding.

N. B. In this part of the Riding, the cuſtomary acre is generally uſed; which contains 7840 ſquare yards.

FARM, No 3. *in the centre of the Riding.*

Soil, red greet, and water ſhaken, incumbent on clay.
Extent, 200 ſtatute acres.

Crops for one year.

43 acres, wheat being 15 acres after fallow.
15 acres after clover lea.
13 acres after oats.
} ſeed ſown from 2½ to 3 buſhels per acre.

16 acres barley after fallow, 3½ to 4 bushels per acre sown

10 acres oats, 5 bushels seed per acre.

14 acres beans and pease, 3 to 4 bushels seed per acre.

70 acres pasture and meadow.

16 acres clover.

31 acres summer fallow.

The farm is worked by 3 ploughs; and 3 unmarried servants, 2 labourers, and 7 horses are employed.

Farm, No 4.

Extent 300 statute acres, half of which is a poor gravel, in a high situation. About 100 acres are annually sown with corn, and 60 acres are fallowed; the leys upon the high grounds are ploughed, after being pastured 3 years with sheep, and sown with oats, or pease and beans. The 2d year they are fallowed, and sown with wheat or barley without manure, and grass seeds, which are pastured for 2 years, or 3 at the uttermost, and then broke up again. By this mode these grounds are kept in good order, while the whole dung raised on the farm is applied to the lower grounds, which are well adapted for the turnip husbandry. Under this system a greater quantity of corn and grass is raised on the farm, without buying any manure, than when the possessor practised a different rotation, expended L. 60 or a L. 100 annually, in purchasing it from the neighbouring towns; besides the profit received from a valuable flock of sheep.

Farm, No 5. *situated in the neighbourhood of Doncaster.*

Extent, 78 statute acres: 27 acres of which are tithe free. Restricted to ploughing no more than 40 acres, which is a dry gravelly soil. Rent L. 196 per annum.

Crops for one year,

18 acres wheat.

8 acres potatoes.

6 acres of oats.

8 acres of pease, cabbages, &c.

27 acres pasture grass, which is eat by 50 ewes and their lambs, 14 cows and 4 horses.

11 acres meadow loam.

FARM, No 6. *six miles from Doncaster.*

Extent, 139 statute acres. Rent L. 110 per annum, tithe free. Soil, lime stone, clay and moor.

Crops sown for 1796,

Wheat	23 acres
Barley	9
Oats	23
Beans	5
Meadow	12
Fallow	20
Pasture	47

Live stock kept on the farm,

8 Horses
5 Cows
1 Bull
20 Ewes
10 Wedders.

FARM, No 7.

Extent, 116 statute acres. Rent, L. 95 per annum. Soil, lime stone and clay.

Crops for one year,

Wheat	22 acres
Barley	12
Oats	8
Beans	5
Meadow	10

Fallow 13 acres.
Pasture 59

Live stock kept on the farm,

6 Horses
6 Milch cows
4 young beasts
2 young Horses
25 Sheep.

FARM, No 8.

The rotation of crops pursued upon a Marsh-land farm, consisting of 432 acres of arable land. The soil where the principal part of the potatoes are grown, a good warp; the other part on which potatoes are also cultivated, a mixture of warp and sand; the remainder of the land, clay, with a small portion of warp, but too strong to grow potatoes, except about 70 acres, which is tolerably good potatoe land, but at too great a distance from the river. Grass land only sufficient to keep two milch cows, and horses necessary for working the farm: 69 acres of the best warp land, divided into three equal parts. 1st, Fallow, with from 16 to 20 loads of manure per acre; set it with potatoes; after, sow wheat; and then fallow again: 3 acres of the same kind of land, that is liable to be damaged by sparrows, when sown with corn, is set with potatoes every year, with about 10 loads of manure per acre each year. 84 acres of the lighter land is divided in the same manner, one third fallow, with ten loads of manure per acre; set potatoes, and then sow wheat; and fallow again. 42 acres of land, lately an old pasture, divided into three parts; one third flax, then sown with rape, and after they come off, plough and harrow the land three or four times, and lay upon it about 20 loads of manure per acre, which will make it in great condition; after which set potatoes, then sow flax again, and rape after. 150 acres divided into three parts; 1st, fallow; 2d, wheat; 3d, beans drilled at nine inches distance, hand hoed twice at 6s. per acre, fallow again, &c. 80 acres of land that was lately in old grass, divided into four parts: fallow, wheat, beans drilled, and oats; then fallow again, &c. The remaining 4 acres thrown to any of the crops that are likely to fail. Rent 25s. per acre; assessments, 5s. per acre.

Distribution of crops for 1795.

		Acres.	Average prod. of an Acre.
Wheat	—	121	from 3 to 5 quarters.
Beans	—	70	from 3 to 6 quarters.
Oats	—	20	from 6 to 10 quarters.
Flax	—	14	from 45 to 55 stones.
Rapes	—	14	from 4 to 5 quarters.
Potatoes	—	68	from 60 to 100 sacks.
Fallow	—	121	
To be thrown where a crop is likely to fail		4	
		432	

Servants, horses, and cows, kept upon the farm:

4 House servants,
16 Labourers,
26 Horses,
2 Milch cows.

The above is an account of a farm, belonging to the best manager in Marsh-land. We must observe he fallows his land very often, yet he is well paid by his superior crops. The last year (1795) he had 100 sacks per acre off most of his potatoe land, and sold them from 6s. to 12s. per sack, of 14 pecks. All their corn is sold by the quarter, of 8 Winchester bushels, though I believe their measure rather over-runs.

SECT. 4.—*Crops commonly cultivated.*

1*ſt*, *Wheat.*—This valuable grain is cultivated to a great extent, upon all the low land of the diſtrict; and is ſown after fallow or turnips, or clover; ſometimes after peaſe and beans. The latter mode muſt be rare, from the nature of the uſual covenants; although we have found, from experience, that the beſt grain, and often the greateſt quantity per acre, is produced after a crop of drilled beans. At all the markets we attended, hardly any white wheat was preſented for ſale; and our information inclines us to believe that little but red wheat is ſown. From trials which we have made upon clay ſoils, we venture to aſſert, that the white Eſſex will yield 3 buſhels more per acre than the uſual kind of red wheat; but we grant, that the latter is better qualified, from the ſtrength of its roots, for being ſown upon all ſoft or ſandy ſoils, where the plant is in danger of being thrown out by the ſpring froſts.

2*d*, *Rye.*—This is a ſevere crop; and, from its uſually ſelling low, ought not to be ſown on valuable ſoils. No great quantity of it is ſown in the Weſt Riding; and, in our opinion, ſoft linky ſands are moſt proper for this grain.

3*d*, *Barley.*—We believe that double the quantity of land is ſown with wheat in this Riding, than is ſown with barley, and that this preference extends over the greateſt part of the iſland. Barley is a tender grain, eaſily injured by adverſe weather, generally raiſed at greater expence, and an acre of its ſtraw will not produce half ſo much dung as that of a crop of wheat upon the ſame land. It is really ſurpriſing, that the price of barley ſhould, in all ages, have been greatly below that of wheat; whereas the latter is generally raiſed at leſs expence, while the former, especially upon clay soils, is a most precarious crop.

4th, Oats.—The general quality of the oats which we examined during our survey, induced us to think, that little attention was bestowed in procuring proper kinds for seed. They appeared, in general, to be of the Friesland and Siberian sorts, which are usually coarse, husky, and defective in meal. No kind of grain sooner degenerates than oats, and the best farmers find a necessity of procuring changes from other soils, so as the quality may be kept up.

5th, Pease.—The breadth of land sown with pease is not great, and perhaps where this pulse is sown broad cast, as little profit is, upon the whole, afforded to the farmer, as from any article whatever. In wet years they yield only halm or straw; and, in dry seasons, the ground is ruined by the weeds, which then enjoy full possession. We venture to say, that pease should never be sown (unless it is the grey sort, or vetches) without being drilled in rows, with sufficient intervals to admit horse-hoeing. In this case, when mixed with beans which keep them off the ground, and allow free air for filling the crop, they will be found profitable and advantageous.

6th, Beans.—From our inquiries it did not appear that many beans were sown in the West Riding, and these were principally in the eastern parts. They were sown in the broad-cast way, which is pernicious in the extreme, and renders a crop well calculated for cleaning the ground an instrument of its destruction. The drilling of beans is now become common in many parts of the island, and we earnestly recommend its adoption upon all lands where the soil is of a proper depth for carrying this plant. They are, on the whole, when drilled and horse-hoed, nearly as valuable, upon clay soils, as turnips are upon those of a different description.

When beans are drilled, we recommend the intervals to be 24 or 27 inches wide, and where turnips are meant as a complete fallow, about 30 or 32 inches. These admit a small plough drawn by one horse perfectly well, which, with the addition of a hand-hoe, is the cheapest and most effectual way of cleaning these crops.

Horse-hoeing beans and turnips has this advantage, that it is the fault of the farmer if his fields under these crops, in the most adverse seasons, be full of weeds. It is well known that beans, from being an open plant at the root, give opportunity to weeds thriving amongst them, which in dry seasons, will ruin them altogether. By horse-hoeing the intervals at proper periods, and running the hand along the drill, they are constantly kept clean; and a well managed field of them, or turnips, will necessarily be as clean as the same crops in a garden.

7th, *Tares* or *Vetches*.—This pulse is of infinite use to the farmer, either for cutting green for his farm stock, or remaining for seed. It is an important article of farm œconomy to have vetches sown at different times, so as maintenance for horses may always be at command: In dry seasons, the second clover crop will often hardly cut, and without this succedaneum, work cannot be carried on, when it is most absolutely necessary.

Winter Tares are sown in many places, particularly about Sheffield and Rotherham; and are excellent spring food for horses before the clover crops are ready. They are sown from September to the 1st of November, and by being cut in April and May, afford sufficient time to prepare the ground for turnips. As they are found to answer so well, we would recommend the cultivation of them, upon all rich warm soils, the maintenance of horses being at that time particularly expensive.

8th, *Turnips*.—Although the turnip husbandry prevails over a great part of the Riding, yet the proper cultivation of that root is not attended to so carefully as good farming requires. Except by a few individuals, turnips are universally sown broad cast, and most imperfectly cleaned (*h*). We understand that it is not much more than thirty years since they were hoed at all; and that the introduction of this most necessary practice, was principally owing to the indefatigable exertions of that truly patriotic nobleman the late Marquis of Rockingham. It may readily be supposed that a people, who so lately thought hoeing unnecessary, will still think an imperfect hoeing sufficient, which we are sorry to say is too much the case (*i*). Indeed it is only by drilling and horse-hoeing that large fields of turnips can be kept in proper order, at a moderate expence (*k*). We saw some fields very well dressed, and carrying good crops, particularly to the southward of Wakefield; but the greater number were full of weeds, in some places too thick, in others very blanky, and not be considered as half a crop, where the management of turnips is well understood.

In order that drilling of turnips and horse-hoeing may be generally practised, we presume that no method could be more effectually taken, than for proprietors to refuse taking broad-cast ones as a fallow crop. It is a mock upon fallow, to consider some of the crops we examined as such; and we are confident, that unless a very great expence is laid out, a broad-cast crop will never allow the ground to be cleaned in a manner equal to where they are horse and hand hoed.

When drilled turnips are meant instead of a complete summer fallow, the intervals ought to be at least 32 inches; and, in this way, if due care be taken to use the hand hoe, the ground will be cleaned in the most perfect manner.

9th, *Potatoes.*—This useful root so beneficial, to the whole community, is raised to a considerable extent in the eastern parts of the Riding, and less or more over the whole of it. They are generally of the kidney kind, although some of the other varieties are as valuable. The same mode of culture will answer for potatoes, as we have mentioned for turnips; and, we need only add, that the drier the soil, so much more will this root be found healthy and nutricious.

Large quantities of potatoes are sent by water carriage, from Selby and other parts of the river Ouse, to the London market; although this root is not a favourite with most of farmers, being a bulky commodity, and yielding little dung, yet, considering the matter in a public point of view, their cultivation cannot be too warmly recommended.

The following account of the potatoe husbandry in Marsh land, we have received from an intelligent gentleman:

Land that is intended for potatoes, if wheat or bean or oat stubble, should be ploughed before Christmass, or as soon after it as possible; about the middle of April, if the land has got well dryed, you must harrow it well, and repeat the harrowing, also use the roller until you have got the land fine. In a few days it must be ploughed again, harrowed and rolled as before; and, if the land be in bad condition, it will be necessary to plough it once or twice more, and work it in proportion; let it lie two or three days betwixt each plowing, and then you may begin to ridge it, plough a furrow round the land down, after which take a breadth sufficient to make a ridge, which should be from two feet eight inches, to three feet distant, according to the state of the land, as fat land requires the ridges to be larger, then when exhausted. If any manure is intended to be laid upon the land, it is usually done in the ridges, and a man with a fork assists in disposing it regularly in the rows. The potatoes are then set upon the manure, and covered with the plough; when the weeds begin to grow, run a plough betwixt every ridge, to cut what may have come up there, and a harrow, trailed by one horse, with the teeth upwards, follows; a day or two afterwards, the horse must walk betwixt the ridges, which it will nearly level, give a great check to the weeds, and warm the land. In a short time the potatoes will make their appearance, and if the land is foul, it will be necessary to use the horse-hoe to stop the progress of the weeds, and give the young plant an opportunity of getting out of their way. When the tops are nearly high enough to ridge up the last time, let the hand-hoes go over them, and cut up what weeds have been left, let them lie a few days, and then begin to ridge them up. The plough should go up and down betwixt every ridge to divide the earth equally, and throw it well up to the roots of the plants, and leave them as near as possible in the middle of the ridges. In about three weeks, if any more make their appearance, pluck up by the hand. When your potatoes are fit to take up, plough out every other row, but be careful to get deep enough least you cut them, gather them into carts, and take them into the most convenient place for delivery, make them into a long pye about three yards wide, and raise them as high as they will lie one upon another; cover them well with straw, and about 12 or 14 inches thick of earth, clap the outside till it's smooth and level, which will throw off the rain, and effectually preserve them from frost. If you intend keeping any till late in the spring, pye them only two yards wide; average produce about 60 sacks per acre, each sack containing fourteen pecks. I suppose they grow annually in Marsh land about 12 hundred acres, all of which are ſent to London. Potatoes grow the beſt upon old going land after beans, next oats and then wheat; but upon land that has been lately broke up, they grow the beſt upon a crop of rape, next flax, and then beans, oats, and wheat, as upon old going land. The ſort ſet are the red noſe kidney, which are procured from the neighbourhood of Berwick, each grower buying as many as will plant three or four acres, which will ſupply him with ſets for the remainder of his land. Until they hit upon this plan of changing their ſeed, they were much troubled with the curl.

Expences upon an acre of Potatoes.

Land rent - -	L.	1	5	0
Working and ridging -		1	5	0
6 ſacks of potatoes at 7s.		2	2	0
Cutting Do and ſetting -		0	2	0
Manure and leading -		2	2	0
Howing, weeding and taking up		1	5	0
	L.	8	1	0

Produce of an acre of Potatoes.

60 ſacks at 5s. 6d. per ſack	L.	16	10	0
		8	1	0
	L.	8	9	0

The ſmall that dreſſes out, of what is ſhipped to London, will deliver them, or rather more.

SECT. 5.—*Crops not commonly cultivated.*

Flax—This is a plant which has never been popular in Britain, and notwithſtanding the premiums which have been ſo long beſtowed upon thoſe who raiſed it, the quantity annually ſown, does not appear to be upon the increaſe; many parts of this iſland are naturally fitted for producing it, and none more than that large tract of ground, upon the banks of the Ouſe, ſituated in this Riding. In the neighbourhood of Selby, a conſiderable quantity is annually raiſed, and from the liſt of the claims given in to the clerk of the peace, for the Weſt Riding, it appeared that the parliamentary bounty was claimed, in the year 1793, for no leſs a quantity than 59,000 ſtones. From our own experience (having formerly ſown many acres with flax,) we can ſay with confidence, that, upon a proper ſoil, no other crop will pay the farmer better than flax; and if due pains and attention are beſtowed upon the pulling, watering and ſkutching, flax of as good a quality may be produced at home, as what is imported from Holland, or the Baltic.

The produce of an acre of flax will be from 24 to 40 ſtone averdupois, after it is clean ſkutched. This operation is performed by the hand, in the Weſt Riding, there being no mills erected in that part of the country for this purpoſe. Some of the flax is allowed to ſtand for ſeed, which of courſe renders the flax of leſs value.

We have found inferior ſoils, ſuch as new broken up muirs, as well fitted for raiſing ſeed as others of a better quality; and they have this advantage, that while the rent is but ſmall, the trouble of weeding them is equally trifling. Beſides, ſeed and flax ought never to be attempted together; when the former is intended, the ground ought to be ſown much thinner, ſo as the plant may have ſufficient air to fill the bolls; whereas, when the flax itſelf is conſidered as the object, it ought to be ſown much thicker, to prevent it from forking, and becoming coarſe; we believe a neglect of theſe things has contributed to render this valuable and neceſſary plant, not ſo profitable as might, from the public ſupport beſtowed upon it, have been expected.

The following intelligent paper on flax huſbandry has been obligingly communicated to us.

The bounty paid for flax and hemp, grown in the Weſt Riding, for the year 1794, amounted to the ſum of L. 720, which at 4d. per ſtone, will make 43,000 ſtone; and taking the average of the crop at 30 ſtone per acre, will give 1410 acres ſown; and from the ſame calculation there would be, in the year 1795, 1650 acres ſown.

As I have not made any particular obſervation on the crops of flax in any part of the Weſt Riding, except Marſh land, I cannot ſay poſitively, what is the beſt ſoil for it. In Marſh land they are allowed to grow as many ſtone per acre, as any part of the Weſt Riding, but not ſo good in quality. Flax, if not ſown upon graſs land new ploughed up, generally ſucceeds a crop of oats; but latterly they have ſown it after a crop of potatoes, upon land that has a few years before been broke up from graſs, and with good ſucceſs. Land that is intended for flax, if an old paſture or meadow land, ſhould be plowed before Chriſtmas; if wheat or oat ſtubble, betwixt Chriſtmas and Candlemas, and as ſoon as it has got well dried in the ſpring, work it with harrows and the roller, till you have got it well pulverized; let it remain in that ſtate for ten days or a fortnight, then open the land out with a harrow, and let the ſeedſman immediately follow. Endeavour if poſſible, to ſow after a ſhower of rain, but wait a few days longer, if the ſeaſon is not too far advanced, rather than ſow when your land is too dry. The rent, if let to a flax grower, is generally from L. 3:10s. to L. 6 per acre.

Hame ſeed is for the moſt part ſown when intended

for white flax; if for feed the Baltic, which makes very good feed next year for white flax, for three or four years after, but must then be renewed. The quantity sown per acre, if for feed, is 8 pecks; if for white flax, from 8 to 10 pecks.

The produce of flax per acre is very uncertain, being a crop that depends so much on a good or bad season; in general from 30 to 50 stones per acre. I have had 70 stones grown; and, from a bad season, I have seen the crop not worth reaping. The quantity of feed produced per acre is from 8 to 16 bushels. I have known 16 bushels of feed, and upwards of 40 stone of flax from the same acre, but look upon 12 bushels of feed, and 30 stones per acre, to be about the average, if the season has been a favourable one.

I do think a good part of the West Riding adapted to the growth of flax, and also that the culture of it has of late been considerably extended. From my own experience, I am convinced that flax is not an impoverishing crop*; for it is generally reaped the latter end of July, which enables the farmer to make a good fallow of his land, and the crop that succeeds it, whether wheat or spring corn, seldom, if ever fails.

Flax, if sown upon good grass land plowed up the Martinmas before, should be in the ground by the second week in April, if the season will admit of it. The feedsman should be very careful to distribute the feed as regularly as possible, it must then be harrowed, two harrows in a place, and one the contrary way after; and if likely to be dry weather should be immediately rolled down. The feed in a short time will be up, and should the season prove favourable, will be fit to weed by the

* We must differ in this matter with our intelligent correspondent, as we have always found flax a very scourging crop.

middle of May, which muſt be done with ſome attention, as much depends upon keeping the land clean to produce a good crop. By the latter end of July, when the the leaves begin to fall off about half way up, and the ſtalks become a pale yellow, it is ready to pull. The work is performed by women at 1s. per day, and the flax, tied up in beats or ſheaves, is carried to the pit, where a man, who is accuſtomed to the buſineſs, puts it in as carefully and even as poſſible, beginning the firſt row with the root-end uppermoſt, but all the reſt with the top upwards; ſo that when the pit is finiſhed, nothing but the top is to be ſeen. Another man covers it with earth, about 2 or 3 inches thick, after which it will require three or four men to tread it night and morning for 5 or 6 days; it will then begin to fall in the pit, and one man will be ſufficient to keep an eye over it, and take care that none be expoſed to the weather, as it will turn black, and conſequently injure its ſale. As ſoon as the baſt or ſkin will peel off readily, from one end of the ſtalk to the other, the ſtalk itſelf break as if rotten, and be a deep yellow, you may then venture to pull it out. The operation is performed with drags, and the flax laid ſtraight and carefully by the pit ſide, where it ſhould remain half a day or more to dry a little before ſpreading. You now take it to ſome land that lately has been cleared from hay, where a man with a proper number of women, (at 1s. per day) attend to ſpread it. The man with a fork gives them the beats or ſheaves as they want them, and takes care that they ſpread it regularly and without lumps, as whatever is left in that manner will turn green and never come to a good colour. After a ſhower or two of rain it muſt be turned, and when the colour becomes bright and even, and the ſkin riſes from the ſtalk, you may venture to take it up. Keep the flax ſtraight and the roots all one way, carry it to your barn, or stack it, if more convenient. In winter dress it out, make it into half stones, and when you have got a sufficient quantity out, send it to market. The weight generally given is 7lb. 2 oz. for half a stone.

Expences upon an acre of flax.

	L.	s.	d.
Seed	L. 1	1	0
Working land	0	16	0
Soding and weeding	0	5	0
Leading, dikeing, &c.	0	10	0
Taking out and spreading	0	12	0
Turning and taking up	0	5	0
Rent of land if let to a flax grower	5	5	0
Dressing 50 stone at 1s. 6d. per stone	3	15	0
Pulling	0	10	0
Profit	7	11	0
	L. 20	10	0

50 stones of flax at 8s. 6d. is L. 20 : 10.

Rape.

It did not appear to us, that rape was much cultivated in any part of the West Riding; and it is only on the eastern parts* that any quantity is sown at all. It is raised both for feeding sheep, and upon account of the value of the seed; although we apprehend, in the last case, it will be found a very scourging crop. There are two ways in which it is consumed by sheep: first, by sowing it in July, and feeding it off before winter; and again in the spring, in which method it is an excellent preparation for barley: 2dly, it is sown upon the wheat stubbles that are intended for turnips (*l*). The land, in this case, is ploughed as soon as the wheat crop is got off, which is usually before the end of August, and it is eaten in spring, previous to working the turnip land.

Both these modes are excellent, and deserve imitation. When rape is intended for seed, it is sown about the 1st of August, either upon fresh land, or land fallowed and dunged. It is cut in the month of July thereafter, by which means it remains near a whole year on the ground.

When waste lands are taken in, they are sometimes sown, after being pared and burned, with rape seed. The produce may be from 2 to 5 quarters per acre, generally 4 quarters; expence of reaping and threshing about 20s. per acre, if stacked and threshed in winter; but, according to the general practice, it is impossible to calculate the expence, the whole neighbourhood being gathered to the threshing, when it is done in the field. In this mode it is a perfect feast, where all comers are welcome; but this *good old custom* is fast going out; and the thriftier practice of stacking it in the yard, and threshing in the winter, introduced in its place. The straw of the rape is sold to the soap boilers at about 5s. per acre.

Liquorice.

WE received the following information from Mr Halley, seedsman and nurseryman at Pontefract, concerning the cultivation of Liquorice. 'The soil most proper for liquorice is that of a dead, light, sandy loam. It is trenched three feet, well dunged, and planted with stocks and runners in the months of February and March, on beds of one yard wide, thrown up in ridges, with alleys between them, and the beds hoed and hand-weeded. The first year a crop of onions is taken in the alleys, and the tops of the liquorice cut over every year. The ground is trenched when the liquorice is taken up, and all the fibres cut off. A considerable quantity, more than 100 acres, is cultivated in this neighbourhood. It is a very precarious plant, often rotten by wetness, and 'also hurt by sharp frosts in the spring and dry weather 'afterwards. Rent of the land upon which it is cultiva-'ted, about 3l per acre.'

Mr Haley also cultivates rhubarb, and has done it to advantage. The quality is esteemed good, and he lately received a medal from the Society of Arts for the cultivation of it.

Woad.

Woad for dyers is raised in the neighbourhood of Selby, among red clover. When it is in full bloom, it is pulled by women and boys, who go before the mowers. It is placed in small heaps, with the tops uppermost; and when completely dried, is put into the barn, and sold to the dyers from 15d. to 3s. per stone. Woad grows well on all lands fit for turnips, and is sometimes taken by itself as a crop.

Sowing Clover for Seed.

Clover being less sown in the West Riding, than in many other districts, it is our intention to confine ourselves here to the method of managing that valuable plant, when the sowing of seed is intended; we therefore class it amongst those crops not commonly sown.

Clover is generally sown in March or April amongst the barley crop, and sometimes amongst the winter wheat, which, in our opinion, will give the greater return; when it is only to remain for one summer, from 8 to 14lb. is sown per acre, which is usually covered in with a light roller. The crop next spring is eat by horses or sheep, and they ought to be removed when rain falls after the middle of May, and the coarse places not close eat, should be immediately cut over with a scyth, so as the next growth may be equal. Five hundred weight of red, and three hundred weight of white, is thought to be an average produce, which, from the prices this article has of late years been ſold at, will yield the grower a handſome return.

At the ſame time it is obvious the ſaving of clover ſeed, in our precarious climate, muſt be a troubleſome proceſs, and attended with conſiderable expence. We alſo ſuſpect, that a clover crop, when the ſeed is ſaved, muſt be a ſcourger; at leaſt we are certain, that Rye graſs when allowed to ſtand for ſeed, will impoveriſh the ground, as much as a crop of oats. We do not ſtate theſe things with a view to diſcourage the farmer from ſaving theſe ſeeds, but only as a caution for him, not to expect that his ground is to be meliorated in the like manner, as if the graſs was cut at an earlier period, or conſumed by cattle or ſheep.

We are of opinion, that the threſhing machine would anſwer well for ſeparating clover ſeed from the huſk, which has hitherto been a difficult buſineſs. If the feeding rollers were ſet very cloſe, we think the ſeparation would be accompliſhed in the moſt effectual manner, or if any paſſed the machine untouched, it could eaſily be put through a ſecond time.

We learn, that Mr Richard Parkinſon, at Doncaſter, laſt ſeaſon, tried garden peaſe, early cabbages, &c. upon one of his fields adjoining that town, and that in general the different crops turned out well. They were managed in the ſame way as we have recommended for beans and turnips; but not being furniſhed with particulars, we are ſorry we cannot detail the exact reſult of his experiment.

NOTES on Chap. 7.

(a) The produce of these lands would infallibly be doubled to the community in value, by inclosing both commons and common fields. Inclosures ought to have been promoted by all the might of the legislature, and if more of this is not speedily done, by removing all impediments to so necessary and natural a work, *famine* and misery of all kinds will inevitably be the consequence; a just and merited punishment for our neglect of the domestic cultivation of our own *bread plant*, and a foolish predilection for the culture of the foreign sugar cane. *W. P.*

(b) This would tend greatly to improve the quality of the stock kept upon them, as the occupier would be enabled to proportion the quantity he put on the land, to the quality of the grass. The proprietor would also have an opportunity of planting the most barren spots, which, in a few years, would contribute to the improvement of the other parts, and afford shelter to the stock. *T. H.*

Section 2.

(c) Good management. *T. York, Esq.*

(d) Is it not better husbandry to burn the couch and weeds, and distribute the ashes upon land? *W. Fox.*

Answer.—Upon many fields, so much of the soil adheres to the couch, that it is impossible to burn it. This is the case with every field, which is here recommended to be summer fallowed. *R. B.*

(e) A few years ago, I was desirous of sowing down a piece of strong land to graze for a conveniency, and could not wait for a summer fallow; I had three crops from it: I ploughed it before winter, then again early in the spring, and the season being favourable, I harrowed, and worked it as well as I could, and picked all the quickens out of it by the hand, at the expence of about 10s. per acre. It was subject to kecks and thistles of different sorts, but knowing these would not all grow among grass, I manured it well, and sowed it with beans and seeds of different sorts, and it is now, and ever has been, as fine a swarth as I have in my farm. I could not hoe the beans, but I hand weeded them, and had a very good crop, when the field adjoining it, from which it was newly incloſed, lay as dead fallow.

A Yorkſhire Farmer.

Section 3.

(? (g) Cabbages or Ruta Bagga, (Swediſh turnips) might probably be introduced in place of turnips, in the fallows of wet or clay lands with advantage. They may be eaten off late in ſpring, when the land is ſufficiently dry to be entered, and preſſed without damage, and if it would not be too late, even in the beginning of May, to ſow the quick growing kind of peaſe, or even barley, either of which followed by clover and wheat, in the courſe would make an excellent rotation for almoſt all lands too wet or ſtiff for turnips, viz. Ruta Baga, peaſe, or barley, or oats, clover, wheat. The connection of the plough with the maintenance of large flocks of cattle, ſhould never be loſt ſight of, ſince a farm, under the plough, will ſupport no leſs live ſtock, than the ſame under graſs; a mighty advantage of the turnip ſyſtem.

" Agricola incurvo terram dimovet aratro,
" Sufficit hinc armenta boum meritoſque juvencos." *Tr. P.*

Anſwer.—We have noticed attempts to introduce a ſyſtem, ſimilar to the one here recommended by Mr P., but they were never attended with advantage. We are decidedly of opinion, that the crop, after a dead fallow, will be of greater value than both the cabbages, and any crop that can ſucceed them; a field of clay land, tread with ſheep to the firſt of May, would turn up in ſo unkindly a manner, that half a crop of barley, could not reaſonably be expected. *R. B.*

Section 4.

(h) There is certainly a great defect in our turnip huſbandry, particularly in the open fields; without making a clean fallow, manuring ſufficiently, and plenty of hoeing, few lands will bring them to perfection. Turnips, without doubt, are a moſt profitable root, when ſown upon ſuitable land, a diviſion of the open fields would facilitate their cultivation. *T. H.*

(*i*) This can only be the case with a few stupid mortals, who never see any persons management but their own; those who look about them, must make the proper distinction betwixt a crop of good turnips and bad ones, which, in the long run, will be found of the utmost importance to the farmer. *A Yorkshire Farmer.*

(*k*) A greater crop of turnips may be obtained by broad-cast sowing, and hand-hoeing, than by drilling and horse-hoeing, as the ground will be more equally planted, and as well cleaned at nearly the same expence. *S. Berks, Esq.*

Answer.—Both these assertions are positively denied; no broad-cast crop, can be so regularly planted as a drilled one, nor so well and so cheap cleaned. *G. R.*

(*l*) This is certainly an excellent method where the ground is rich enough to grow rape; but I believe, few soils, except the rich warp and loamy sands, are capable of doing it. *T. H.*

CHAPTER VIII.

GRASS.

SECT. 1.—*Natural Meadows and Pastures.*

IF by the term meadows we were to underſtand only ſuch fields as are occaſionally overflowed with water, and unfit for cultivation, a very ſmall portion of the diſtrict would be claſſed under that head; but as the old paſtures are generally diſtinguiſhed by the name, although in our humble opinion, very improperly, we are under the neceſſity of including both in one ſection.

The old paſture lands of Yorkſhire have remained in that ſtate for a long ſpace of time, probably ſince the incloſures were made; and unleſs upon particular ſoils, naturally adapted for graſs, their value cannot thereby be increaſed; but on the contrary, when incumbent on clay, till, or limeſtone, they turn ſour, full of bad plants, and are proportionably late in their growth, which renders them leſs valuable to the poſſeſſor.

This deſcription will apply to a conſiderable portion of the Weſt Riding, and from the vigilance with which it is preſerved in this unimproving and non-productive ſtate, a ſtranger would be apt to believe that eſtates were entailed with that burthen upon them. As this excluſive ſyſtem is, in our opinion, detrimental to the public, we ſhall attempt to ſhow, that breaking up theſe grounds could in no ſhape hurt the proprietor, but on the contrary would materially promote his intereſt (a).

Does ploughing the ground in a proper manner reduce the natural value of the ſoil? or, in other words, will it hinder land from carrying graſs of good quality when it is laid down again? So far from that, it is often neceſſary to convert paſture into tillage, merely, that better crops of graſs may be afterwards produced. Land, when uniformily kept in one courſe, tires for want of variety; and a farmer might as well expect his land to carry good wheat every year, by the force of manure, as look for graſs of equal value for a continued ſpace of time. It is found that the two firſt years of graſs, when the land is ſown properly, afford a greater return than the ſame number of ſubſequent years. The crop is conſiderably earlier, therefore of greater value; and, from the natural vigour of the plants, a large additional quantity of paſture is procured (b).

But allowing, for argument ſake, that the land when in graſs continues in a progreſſive ſtate of improvement, ſtill a conſiderable ſum is loſt to the proprietor from not ploughing his fields. We hold, that land, after it has lain a certain number of years in graſs, is able to pay an extra-rent. This, by continuing it in the ſame ſtate, is totally loſt; becauſe if it were ploughed for ſome years, and then ſown down and clean in good heart, it would carry more graſs than ever.

A very great loſs is ſuſtained by the public from the practice of this excluſive ſyſtem. It requires no figures to ſhew, that by breaking up land, at proper intervals, a great deal more corn would be raiſed, an additional quantity of manure procured for enriching barren ſoils, and much employment conſequently given to the people at large. Theſe are important matters, and ſhould be ſeriouſly weighed by every proprietor who keeps his estate principally in grass.

It may be asked, if the grass grounds are broken up, how are cattle to be fed for supplying the butcher? We answer, by laying down the old ploughed fields, which would be as much benefited by a cessation from ploughing, as the others would be renovated by it. We apprehend as much grass would be raised, in the way we are describing, as ever, while at same time the quantity of corn would be greatly increased.

With regard to the western parts of the Riding, where there is at present nothing but grass, we are dubious whether we can recommend cultivation by the plough in the same extent. The climate is wet, and corn husbandry must be precarious. But we are convinced of the propriety of raising as much as is necessary for supporting the inhabitants. Corn has already been cultivated there, for all the low fields have at one time or other been ploughed; and we suppose, the climate would then be similar to what it is at present. We have no doubt, but that by sowing grain very early, it might all be harvested in proper time. Fallow wheat might be sown by the end of August, or first of September, which with Dutch, or Poland oats, would always make an early harvest. But before any of these rich fields can be broke up, the tythe system must undergo a change, as it would be a notable affair for a tithe-holder to have a tenth of the weighty crops they would produce. From respectable authority we learned, that the payment of tithes, was in a great measure the cause of laying these fields totally in grass, and that this tax continues to operate as a prohibitory restriction against breaking them up.

A few fields which may properly be distinguished as meadows, are scattered up and down the Riding, but they are not of great importance. Draining is the first step to improving them, but as they are generally common, this can hardly be attained without a previous division. It is rare that many people can agree concerning the necessity of making, or the mode of executing improvements, and this furnishes the strongest reason for all land being held in severally, which gives full scope to ingenuity and enterprise.

As low grounds adjoining to rivers must naturally accumulate the sediment of water brought from the higher lands, so we find that the greatest part of the ground, on the banks of the Ouse, of a rich quality, producing great crops both of corn and grass. That tract of ground called Marshland, has at one time or other, in all probability, been totally under water, as the surface is generally of that sort which obtains, in many parts of the island, the name of *water-fat* soil.

SECT. 2.—*Artificial Grasses.*

THE grasses that are cultivated are red clover, when it is to be followed with wheat, and white clover and hay seeds for pasture. Sometimes hay seeds are sown by themselves, and a good deal of Sainfoin is cultivated in the neighbourhood of Tadcaster and Ferrybridge. As for the old rich pastures about Skipton, Settle, and other places, it is not easy to say what they have originally been sown with (*c*). There appears among other grasses, a great quantity of what is called honeysuckle grass, which we suppose to be the same plant sold under the name of *cow-grass* by the London seedsmen. Most of the vale of Skipton has been 50 years in the same situation as at present; and the proprietors do not seem anxious for changing it (*d*). The quantity of hay feeds fown upon an acre is very great; no lefs than three quarters. Probably fome people may fow lefs; but we had accounts from fome very judicious farmers that the above, when fown with 18lbs. white Dutch clover, afforded them the beft pafture. Indeed none of them can fay what thefe hay feeds are; they may be weeds or other noxious trumpery; this they could not explain.

There is very little rye-grafs fown. The people in general have a mortal averfion to it; and the clover crops (*e*), from a want of this mixture, make exceeding bad hay (*f*). The old paftures are therefore frequently cut, which makes a hay of great repute, and is generally ufed over the whole Riding.

As it is thought neceffary to invigorate thefe old paftures with dung, after being cut for hay, we prefume it would be fully more advantageous for the occupiers to refrain from taking this crop, and to confine the ufing of their dung, to lands which are in a ftate of tillage. Under the prefent fyftem, we confefs this rule would be improper; but upon the fuppofition that the old grafs lands were broken up, and brought into a regular courfe of cropping, we earneftly recommend, when land is fown down with an intention to be depaftured with cattle or fheep, that the fcythe fhould never be admitted into fuch fields, unlefs to deftroy thiftles or other weeds.

The quality and kind of hay feeds, generally ufed, when land is fown down for pafture, is not eafily afcertained; for the very fowers of them, in moft cafes, are abfolutely ignorant of their properties. To us it appears they are fown to a wanton and unneceffary extent, and that good pafture could be got from fowing graffes of other forts, the qualities of which are better known, and which would be easier eradicated when the ground is broke up for tillage.

The grasses that in our opinion are most profitable to the farmer for pasture are, white clover, tresoil, and rye grass; perhaps where sheep pasture is intended, a small quantity of rib grass is not improper. The quantities of the above seeds that we recommend for making a good and close bite, are, 12 lb. white clover, 12 lb. tresoil, and one bushel of well cleaned rye grass, for a statute acre. We are much mistaken if these will not at once fully cover the ground, and from their springing at different periods, fresh grass is always afforded to the flock. The expence of sowing an acre in this way, will upon an average of prices, be from 16s. to 18s.

Where grass is intended for a hay crop, very different management is required. In this case, thick sowing weakens the plants, and deprives them of their vigour and strength: 14 lb. of red or broad clover, and half a bushel of rye grass, is perfectly sufficient; and, with these quantities, we have often seen as strong grass as could stand. Clover, by itself, always makes bad hay, although we are ready to acknowledge, that rye grass is detrimental, if wheat is intended to succeed. But considering the clover as a crop intended for eating green, or for making hay, there is a necessity for giving it a body and strength, by a small intermixture of rye grass, and the above quantity is sufficient.

It remains to mention that wherever grass seeds are sown, it is indispensably necessary that the ground be in a proper state of culture, and reduced as fine and equal as possible, or else the one half of the seeds will be lost. For want of attending to these precautions, great loss is often sustained, as not only the crops of grass are rendered small and scanty, but a failure in this respect is detrimental to the succeeding rotation.

SECT. 3.—*Hay harvest.*

The hay harvest of this district is regulated by the soil, climate, and age of the grass, which is to be harvested, and continues from the middle of June, to the end of August. A principal object which ought to be attended to is, never to cut grass during rain, which increases trouble during the remaining stages of the work, and often proves injurious to the quality of the crop. Perhaps the best method of winning clover hay is to let it lye, when the weather is dry, in the swath for twenty-four hours, then turn or shake it as circumstances require, and to put it up in small cocks immediately, or during the course of the day. These cocks ought to be gathered into large ones, as fast as the natural moisture evaporates, which, if properly built, will preserve it from danger till it is in sufficient condition for stacking.

The process of making hay from meadow grass, is necessarily more difficult, and in many seasons, the crop can hardly be saved under every exertion. The difficulties naturally attending this sort of hay, are increased by the smallness of the inclosures, which retard the free circuation of air. In a word, the hay harvest of this district is at all times a troublesome, and, in bad seasons, an expensive process.

SECT. 4.—*Grazing.*

The West Riding may be considered as a great feeding district, and the graziers in general are very expert at their business. Horned cattle of all kinds are here fattened in a complete manner, the best evidence of which is the quality of beef and mutton offered to sale in all the public markets. The district is neither able to supply its consumption with lean cattle, nor sheep, as immense quantities of both are annually brought from Scotland, and the contiguous northern counties.

Cattle are generally made, what in many places would be called fat, upon grass, and afterwards finished by stall feeding with turnips, or allowed to run in the small well sheltered closes, and turnips or hay carried thither for their food. The first way of feeding we consider as the best. It is most thrifty, affords a greater quantity of dung, saves the ground from injury during bad weather, and rots the straw used for littering the cattle.

Sheep are sometimes fed off upon the turnip field, a small part of it being inclosed with hurdles; from any observation we could make, this did not appear a general practice, although in our opinion it is the most profitable way of consuming turnips upon light dry soils.

Mr Stockdale, at Knaresborough, writes us, that in the year 1793 he had fed three oxen upon lintseed cake, &c. which had been wrought the former year, and gives us the following statement of his profit:

Debtor.

	L.	s.	d.
To cash paid for three oxen -	L. 19	10	0
To summer feed - -	7	17	6
To 10 weeks on after-math, at 12s. per week	6	0	0
To 2 tons lintseed-cake, with freight and carriage to Knaresborough -	18	0	0
To hay and attendance -	7	0	0
To oat meal and hay for ten days previous to their slaughter - -	1	10	0
	L. 59	17	6

		Brought over,	L. 59	17	6
To neat profit	-	-	30	2	6
			L. 90	0	0

Creditor.

By Cash for three oxen	-	-	L. 90	0	0

They were killed at Knaresborough, in spring 1793, by Messrs Buckle and Farnell, and never was more blooming, better, or tenderer beef, sold in that market. Their last ten days food had completely corrected the oily quality of their former diet; and it is a known fact, that working beasts fatten quicker than those that have not been inured to labour, and quick fed meat is always most tender."

Mr Parkinson, at Doncaster, writes: "We have a sort of sheep from Northumberland that feeds well, and pays a great deal of money. I had last year 20 ewes from that county: bought them October 1791, put them to a Dishley tup, and kept them on till December, 1792.—Profit as follows:

"Sold the wool for	-	-	L. 5	0	0
27 lambs, at 18s.	-	-	24	6	0
4 ewes, at 50s.	-	-	10	0	0
16 ditto, at 45s.	-	-	36	0	0
			75	6	0
"Prime cost		-	24	10	0
"Profit	-	(g)	L. 50	16	0"

In the western parts of the Riding, a number of hogs are fed upon oat meal, and sold to the Lancashire manufacturers at 7s. per ftone, of 14 lb. avoirdupois; the hams are ufually fent to the London market, as nothing will do with the Lancafhire people, but the fatteft parts of the beaft.

NOTES on Chap. 8.

(*a*) It is certainly a very mistaken opinion, that some old pastures of good land, should not be taken up. There is a time for all things, and no land should lie in grass for ever. I have, in my farm, old pastures which have lain time immemorial, and which are worse by one third, than I have known them within these 20 years or less. I have other lands which have been inclosed from the common fields, and sown down to grass from about the same time, which will feed twice the quantity of stock, as the old pastures here alluded to, though not equal in quality. This land has continued in good condition, by pasturing it with sheep and other stock, and may do so for some time, till it becomes mossy and hide bound, and then it should certainly be pared and burned, as should all lands in this condition. But the stewards and landlords are averse, to these old pastures being disturbed, notwithstanding the advantage which might be derived to the tenant, as well as the increase of manure, and considerable improvement of the land so managed.

Two, three, or four years is sufficient time for this land to lie in grass, and no land should lie more than ten or fifteen years, if the benefit of the farmer is considered, as well as that of the landlord. *A Yorkshire Farmer.*

Taking up old grass land, and laying down the old tillage fields, I consider as material improvements. *S. Birks, Esq.*

(*b*) Not a doubt about it, in many places plowing old lays, and laying down new ones, are essential to the landlords, as well as the farmers interest. Variety is charming even to old mother Terra. *W. P.*

(*c*) It is pretty certain, that the pastures about Skipton and Settle, at no time have undergone the discipline of a regular course of tillage; probably many of them might be improved by being broken up, thoroughly worked, cleaned, and skillfully laid down with the finest hay seeds, before the richess of the soil could be exhausted. Such an undertaking should be executed by the land owner; it would be dangerous to trust the power to a common farmer: if it were granted without very strict covenants, a dreadful havock might be expected; however, few of the proprietors would easily be induced, to make any experiment with the plough, upon these rich pastures, and a doubt cannot be entertained, whether their value would be very greatly reduced, by converting them permanently into arable lands. *T. York, Esq.*

Answer.—It is truly astonishing, to hear any person in his senses, speak in the above manner; if good land was to receive the injury he dreads from the plough, the greatest part of the land in the kingdom, would have long ago been reduced to a *caput mortuum*. The writer of this answer has often plowed old grass fields, without being fettered by the strict covenants, recommended by Mr York; and he can, with confidence, say, that after being cropped for a number of years, they carry him as good crops as ever. He thinks if leases of a proper duration were granted, (he means much longer than 21 years), that land would never be exhausted; but short leases require both the spur and the whip, and so the land is at last injured. *R. B.*

(*d*) The proprietors there are justly afraid of the plough, with its blessed companion the tithe waggon. They will not suffer the tenants to plough an acre, though, from the want of straw, and the very high price of bread corn, from distant carriages, there is every encouragement to grow it, but the tithe.

A Yorkshire Freeholder.

(*e*) We do not find this to be the case. Clover hay is a much richer food by itself, than when mixed with rye grass. *T. H.*

(*f*) The best in the world almost, in *dry* seasons, for winning; and the worst, in a *bad* one. *W. P.*

(*g*) I should be very sorry to contradict Mr Parkinson's calculation; but I will venture to say, that no other person can make the like profit by 20 ewes. *A. Farmer.*

CHAPTER IX.

GARDENS AND ORCHARDS.

IT is perhaps out of our province to enter upon these articles, unless it be so far as respects the kitchen garden. Every farmer ought to have a piece of ground adjoining to his house, properly fenced, for raising pot herbs and other vegetables; and it would be advantageous that this was of such a size as to admit cultivation by the plough. The expences of a garden, in other cases, is often more than the value of the produce, and we see no cause why vegetables of all kinds may not be raised in a garden, according to this method, as well as potatoes, beans, and turnips, which are cultivated in the fields.

Every cottager ought likewise to have a piece of ground for a garden, upon which he may, at no expence, raise vegetables for his family. We suppose one sixteenth of a statute acre would be fully sufficient, and this he could easily cultivate by hand labour, during his leisure hours.

A particular species of plum grows at Sherborne, and in the neighbourhood, called the Winesour. It grows well, both upon gravel and lime-stone, is hardy, a good bearer, and answers upon any soil; but does not bear so well, nor its flavour so good on any as on lime-stone or gravel. On a strong deep land, the trees run too much to wood, and do not bear fruit in proportion. These plums blosſom better than any other ſort, and are produced from ſuckers. The fruit ſells from 21s. per peck, when ſound and good, to 4s. 6d. when cracked and damaged. They are eaſily hurt by rain.

CHAPTER X.

WOODS AND PLANTATIONS.

THERE is a great deal of oak and ash wood grown in the West Riding, which meets with a ready market at the shipping and manufacturing towns. Much is also used at the mines and coalieries. The Duke of Norfolk has above 1500 acres of wood land in the parish of Sheffield, and we believe great attention is paid, both by him and other proprietors, to the management of this valuable article. Large quantities of logs and deals are imported from the Baltic, which, at a future period, might be unnecessary, if Scots fir, and larches, were planted upon the waste grounds.

Messrs Tweedale and Noble, stewards to Mr Beaumont of Brettonhall, who possesses a great deal of valuable timber, say,

" It is the custom of the country, when a wood is ready to go down, to set out and leave as follows, viz.

Every 21 years

Poles, supposed to be left for a future fall, being judged to be 20 years old, which, in 40 years more, it is supposed, would be timber trees,—left on an acre upon an average - - - - 180

Trees, supposed to be 40 years old, left on an average per acre - - - 10

Timber trees, supposed to be 60 years old at the time they are ready to go down, judged to be taken down on an average per acre - 10

Reaſons why the underwood is not kept cut quite down, viz.

"The bruſh or underwood would not turn to any profit, except that it ſtands for 21 years, and then it is taken down along with the timber, for different uſes; ſuch as binding hedges, making riddles, burning for charcoal, and many other uſes. The trees that are left are at ſuch a diſtance from one another, that they do not prevent any thing from growing, but what will pay in twenty years time: but if the bruſh or under wood was kept quite cut down, it would neither be ſo well for the timber and younger wood; that method having been tried, it was found that neither the wood nor the bark made ſo much improvement, owing to its being ſtarved in the bottom, when the underwood was not admitted to grow.

"A tree left for a future fall, is chiefly one that grows from its own ſtem, and what we call a lording, and perhaps only forty years old, which, to ſtand twenty years more, in general pays better than to take it down at that age.

"It is ſuppoſed, when a fall of wood is ready to go down, that with the poles, underwood, &c. it is worth fifty-five pounds per acre, upon an average.

"The value of wood ſet out to ſtand for a future fall, is judged, at the time of its being left, to be, upon an average, worth eighteen pounds per acre.

"The woodlands in general, if they ſhould be quite cleared of all the wood, underwood, &c. and put into cultivation, which would be at an enormous expence, it is ſuppoſed, would only, upon an average, be worth five ſhillings per acre.

"It remains to be added, as another reaſon for taking down wood in the manner we do, that by this method we have wood for all ſorts of cuſtomers; and as ſuch can dispose of it to more advantage and convenience. The small wood is used for laths, baskets, puncheons for coal pits, hedge stakes, &c. the larger for husbandry implements of every description; the large timber for house-carpenters, ship-carpenters, coopers, &c. &c."

Much has been said of late, relative to planting the waste grounds, and we concur with those who think such a subject deserves the strictest investigation. Britain, in a great measure, depends on foreign countries for being supplied with timber, while thousands and thousands of acres at home, capable of producing it, are allowed to lye waste, and nearly unproductive. We do not mean to recommend an extension of wood, where the land is of a superior quality; because we are convinced that such soils will pay more under corn and grass, but surely those parts to which nature has not been so liberal in her gifts, cannot be more advantageously employed than in the growth of wood suited to its soil, situation, and climate, and there are very few of the most barren and exposed wastes but what, under proper care and attention, will produce wood of one kind or other.

It cannot be too strongly inculcated, that where a new plantation is to be made, the strictest attention ought to be paid to fence it at the first, in a substantial manner. If young timber is once injured, it never thrives afterwards; and a loss of that nature renders the whole original outlay in a manner useless.

Perhaps it would be of great public utility that an act of the legislature was passed, requiring every landed proprietor to have a certain number of acres of his estate in woodland. If the prosperity of the kingdom be concerned in our having a sufficient quantity of timber; and if the quantity be annually decreasing, as some late writers maintain; then surely it ought to be a material object with every landed gentleman to supply that deficiency, by laying out a certain portion of his estate in the planting of trees.

CHAPTER XI.

WASTE AND UNIMPROVED LANDS.

THE waſte lands, in this diſtrict, are very extenſive, amounting, according to Mr Tuke's calculation, to two hundred and ſixty-five thouſand acres, which are capable of cultivation (*a*), and one hundred and forty thouſand acres, which are incapable of improvement, except by planting; being rather more than one fourth of the whole lands of the diſtrict. If we add to theſe the common fields, which are alſo extenſive and ſuſceptible of as much improvement as the waſtes, it will at once appear how much remains to be done, before the cultivation of the diſtrict can be pronounced finiſhed or perfected.

The quantity of waſte land is diminiſhing every day, as incloſure bills are frequently paſſed for that purpoſe; but ſtill a great deal remains to be done. There are many parts of theſe waſtes capable of great improvement if divided and incloſed. But the far greateſt part would not repay the expence of incloſing; at ſame time it is our opinion, that larches and Scots firs would thrive in many ſituations. (*b*) Wood of theſe kinds is much wanted, and we apprehend, would pay the proprietor well, and contribute to the public convenience. At any rate, as the waſtes are moſtly common, the proportion belonging to each proprietor ought to be aſcertained, which would enable him to improve his ſhare in the manner he might ſee moſt advantageous. (*c*)

The common fields, as is already said, are numerous and extensive, and the husbandry carried on upon them is uniformly bad. They are generally of the best kind of soil; but are worn out with long and successive courses of cropping, which have probably been the same for several centuries. The proprietors of them are mostly sensible of the defects, necessarily accompanying common field management, which must be evident to them from land rising at least one fourth in value, when it is divided and free scope allowed to the genius and talents of the farmer; but the expence of a particular act of division intimidates many from applying to parliament for its interposition. It would therefore be of great utility that a *general bill* was passed for that purpose, as is already the case in Scotland, leaving it to the judge ordinary of the bounds to put it in execution, when application for that purpose was made by any of the proprietors. It would be necessary in this bill to define the extent of manorial rights, and to settle the proportion to be allowed for tithes, in case they are not previously regulated. If the fields are divided, we see no necessity to force the proprietor, to inclose whether he will or not, as is done at present, in consequence of the powers vested in the commissioners appointed to execute the respective inclosure bills. If the proprietor is attentive to his own interest, he will do it himself, without compulsion, and at the same time do it more frugally, than when it is executed under a public commission.

With regard to the waste grounds which are very extensive, they ought to be divided wherever they are common. At present they are of very little profit to the different proprietors, being in general vastly overstocked, unless where they are stinted pastures, which is not frequent (*d*). If each person's proportion was duly ascertained, he could manage his own part as he saw most conducive to his intereſt. If it was worth while, he would incloſe and improve. If it anſwered for planting, he might improve it in that manner; or he would depaſture it with ſuch ſtock as he judged moſt proper and advantageous (e).

As we have mentioned the Scots law for dividing commons, we give the following extract of the act of parliament paſſed in 1695, for regulating that buſineſs.

'All commons, excepting thoſe belonging to the 'king in property, or royal burghs in burgage, may be 'divided at the inſtance of any individual having inter- 'eſt, by ſummons raiſed againſt all perſons concerned, 'before the Lords of Seſſion, who are empowered to diſ- 'cuſs the relevancy, to determine upon the rights and 'intereſts of the parties concerned, to divide the ſame 'amongſt them, and to grant commiſſion for perambu- 'lating, and taking all other neceſſary probation, to be 'reported to the Lords, and the proceſs to be ultimately 'determined by them, declaring, that the intereſt of the 'heritors having right in the common ſhall be eſtimat- 'ed according to the valuation of their reſpective lands 'and properties; and that a portion be adjudged to each 'adjacent heritor in proportion to his property; with 'power to the Lords to divide the moſſes, if any be in 'the common, among the parties having intereſt; or 'in caſe they cannot be conveniently divided, that they 'remain in common, with free iſh and entry, whether 'divided or not.'

Upon this article, Mr Payne at Frickley ſays, "A conſiderable proportion of the arable land is uninclosed, to the great obſtruction of agricultural improvement; the advantages of incloſing are numerous and important. The liberal occupier of *incloſed* land, whoſe mind is actively improved in the employment and increaſe of his produce, with whom innovation has no fault, but when it is useless, this man on *inclosed* land has not the *vis inertiæ* of his stupid neighbour to contend with him, before he can commence any alteration in his management, that he is clearly convinced will be to his advantage; he is completely master of his land, which, in its open state, is scarcely *half his own*. This is strongly evident in the cultivation of turnips, or other vegetables for the winter consumption of cattle; they are constantly cultivated in inclosures, when they are never thought of in the open fields in some parts; and I know no township in this Riding, except that of Wath upon Derne, where the turnips are cultivated in any degree of perfection in open fields. At that place, they have long been wisely unanimous on the management of their common fields, and in selling the whole turnip crop, by a valuation, to a person engaging to stock them entirely with sheep on the land: but even *there* they cannot apply their own produce to the improvement of their *own* stock, nor have they it in their power to vary their management by the introduction of any grasses for more than one crop in their rotation; both essential articles, when the improvement of live stock, particularly sheep, is in contemplation; this argument for inclosure might be very amply dilated on, were I writing a treatise instead of a letter, for it is clearly of importance to the cause.

"Common fields are frequent; the difference of value at present between common field, and inclosed land of similar quality, is about one-third greater in favour of the latter; but if the spirit of improvement was a little more awakened, this difference would be greatly increased.

"There are great tracts of waste land in this neighbourhood; I may extend this remark to the whole county: lands now utterly lost to the community, even in this rich and populous Riding; and be it mentioned to the utter disgrace of every thing in the country, that after a long period of years, in which this island has depended on foreigners for a part of its necessary consumption, these lands are still waste; they are a complete nuisance to every occupier, who has the misfortune to border upon them; whose inclosures are certainly exposed to the inroads of their pining inhabitants, which you scarcely *guess* to be *sheep*, but for the bits of ragged wool they carry on their backs: the *feats of activity* of these animals are such, that no fence can prevent their performing them. These wastes are certainly capable of every improvement by inclosure, which is their *sine qua non*."

We cannot dwell sufficiently upon the happy consequences, which would certainly accompany the enactment of a law, for the general division of the common fields and wastes. The present mode is uncertain, in some cases impracticable, where the lord of the manor, or the tithe holder refuse an agreement for their claims; expensive in an eminent degree, allowing no opposition is made, and even upon the supposition, that an equal number of inclosure bills will pass annually, as has done for an average of twenty years past, which is as much as can be expected, yet still the common fields, and wastes, will not be divided for a couple of centuries. While we earnestly recommend a general division bill, we as decidedly recommend, that the inclosing of lands ought not to be a compulsory measure; many fields will not pay the expence; besides, if the legal obstructions were removed, every indevidual who discerned his own interest, would, in practicable cases, set about that work himself, which he assuredly would perform at far less expence, than when the business is executed under the controul and direction of persons, who are in no shape interested in the success of the measure.

Mr Stockdale at Knaresborough, a gentleman to whom

we are under the greatest obligations, and who deserves the thanks of the Board for his unwearied exertions to facilitate the work, in which we were employed, describes, in a circumstantial manner, the difficulties which attended the division of Knaresborough forest, an extent of ground of no less than 33,000 acres. Here follows his account of that business and the manner in which he improved the share alloted to him.

"The forest of Knaresborough, till the year 1775, consisted of a great extent of ancient inclosed land, comprized within eleven constableries, or hamlets; to which belonged a tract of upwards of 30,000 acres of common, whereon Knaresborough, and several other towns, not within the other constableries, claimed, and had exercised a right of common, and turbary, equally with the owners of property within these eleven constableries. This waste, in its open state, yielded the inhabitants fuel, and pasturage for their sheep, horses, and stock of young cattle; and some opulent yeomanry profited exceedingly thereby; but to the necessitous cottager, and indigent farmer, it was productive of more inconvenience than advantage; if not to themselves, at least to the public at large, who was by that means deprived in a great measure of the exertions of the farmer, and the labour of the cottager and their families; for it afforded their families a little milk, yet they would attempt to keep a horse, and a flock of sheep. The first enabled them to stroll about the country in idleness, and the second, in the course of every three or four years, were so reduced by the rot, and other disasters, that, upon the whole, they yielded no profit.

"In 1770, after various struggles, an act was obtained to divide and inclose this extensive waste, and the powers thereof committed to no less than five commissioners, and three surveyors, all or most of them unequal to the undertaking, from whom both great delay and expence were incurred. After four years had elapsed, an amendment of this act became necessary, which was obtained in 1774. Thereby a sixth commissioner was named, who had been appointed a surveyor by the first act, and who had thought proper to execute his duty by a deputy. In 1775, the commissioners made out a description of their intended allotments; and in or about the year 1776, they executed their award, which unfortunately is deficient in every essential requisite: but with all these inconveniences, the generality of proprietors, to whom allotments were made, and particularly the small ones, set about a spirited line of improvement. The poor cottager and his family exchanged their indolence for active industry, and obtained extravagant wages; and hundreds were induced to offer their labour from distant quarters; labourers of every denomination, carpenters, joiners, smiths, and masons, poured in, and met with constant employment. And though, before the allotments were set out, several riots had happened, the scene was now quite changed; for with all the foreign assistance, labour kept extravagantly high, and the work was executed defectively, and in a few years many inclosures almost prostrate, and of course required making a second time. All these circumstances taken together, were a heavy load upon the allotments, and in general rendered them very dear purchases. The forest, however, got in a great measure cultivated, and rendered a wonderful increase of product to the public, though at the expence of individuals. A public, or turnpike road was opened through the centre of the forest, which afforded an easy communication between Knaresborough and Skipton in Craven, and the manufacturing towns in the north-east of Lancashire. And though scarce a single cart was before seen in the market of Skipton, not less than 200 are weekly attendant on that market at present.

"In consequence the product is increased beyond conception, the rents more than trebled, and population advanced in a very high degree; indeed the lands, both ancient and those newly inclosed, being exonerated from tithe, a full scope was given to spirited cultivation; and to the credit of small proprietors, they took the lead, and brought their small shares first into the completest state of cultivation (*f*). I wish it was in my power to say as much of the large proprietors, but facts will not warrant it. On the contrary, I know of very few men of independent fortune, or others to whom large tracts were either assigned as their stipulated share, or acquired by purchase, under the clause for sale to defray the expence of the act, who have made any improvement, or scarcely effectually ring-fenced their property.

"Many impediments prevented their activity; first, what was to be done must be committed to the care of servants, or agents; secondly, the extravagance of wages, by reason of the want of inhabitants; and above all, the impossibility of letting large tracts as farms, where it must be a series of years before any returns could be expected, or even provision obtained for their working horses. These obstacles operated to a total neglect, or desertion; and in consequence, large tracts indeed at this hour are in their wild uncultivated state.

"If I may be allowed to offer my sentiments how to turn these tracts to better advantage, I should advise building a number of cottages, with suitable small outbuildings, and laying to each not more than 10 acres of land; tempt individuals by suffering them to live rent-free for the first seven years, but obliging them to break up two acres annually, till the whole was improved; then fix a reasonable rent, and add 10 acres more for the same term, and conditions; and so proceed gradually, till the whole of such part, as would admit of cultivation, was gone through. The land thus improved, would be considered by the inhabitants as the work of their own creation, and nothing but cruel treatment by their landlords would drive them away. In a few years population would improve, and that once locally obtained, every other difficulty would vanish.

"Several considerable tracts of this forest have fallen to my lot, both as assignments in right of former property, and by purchase; most of them were of the worst strata, being either confined bogs, or cold steril clay, mixed with white sand, and the surface, pared off for fuel. Little profit could be expected from such kind of property; but nevertheless, I attempted improvements, which many condemned me for; and I frankly confess, my expectations were not gratified, though I still flatter myself my efforts are not wholly useless, as my errors may probably enable others to benefit by shunning the like plan.

"I will state the means I first took, and then point out the errors, or propriety of them; and afterwards give a short account of my present mode of management.

"When I first took possession of the clay parts, so injured as stated, by being pared for fuel, I was eager to get my ring-fences completed, and thereby was led to give extravagant wages, and by employing strangers had them badly executed; these men wanting subsistence-money, while completing a contract, were generally in advance before their labour, and rarely finished them, even in their own defective mode, and the work, particularly stone fences, was to do over again; this was folly. I then purchased oxen to plough with, and ploughed as deep as possible; by which means stones were ploughed up, where none were expected, which would have made the fences, and saved a great deal of the former expence of leading from a distance. Had I now to begin, I should first plough as deep as I could with oxen, collect the stones raised thereby, and make a broad case of a fence, at least 30 inches, and raise the wall no higher than the stones would serve to surround the allotment; and rest satisfied therewith till the next ploughing, whereby more stones would arise, which I would use in raising the wall gradually to its proper height; by this means, the walls would be more substantial, and raised at one-third of the expence.

" After the first deep ploughing, I left it in that state a year, exposed to frost and heat, then harrowed well, and ploughed across, and added three chaldron, or nearly 100 bushels of lime per acre, to make the land fall, and correct the acidity: and in the spring following sowed with oats, after a third ploughing; and the next year, pease or vetches; then fallowed, and limed as before, and took two crops to each fallow so limed, until I found the repetition of lime did harm, instead of being of advantage. In place of this, I now take one crop to each fallow, have better crops, and save two guineas per acre by withholding the lime, which cost me 14s. per chaldron; by this means I get only six crops in twelve years, but which produce more than eight crops by the other mode, keep the land in better condition, and save eight guineas, before expended on lime. Probably lime may be again necessary at a future day; but I am confident, that with some sorts of lime, you may use it till the land will neither produce corn nor grass. The quality of lime varies much; we have two sorts, one burnt near Ferrybridge, and another at and near Knaresborough; where the heaps of the first are laid, there is always the best crop; but where the heaps of the other are laid, you will frequently find the land steril for several years. The firſt ſort is burnt from compact ſtrong ſtone, the other from a porous marly ſtone. At proper intervals I ſow with graſs ſeeds, eat them the firſt year with ſheep, and lay all my fold-yard compoſt on the graſs, except where ſome ſmall parcels that will grow turnips demand it. But this kind of clay land will neither anſwer well long in tillage, nor in graſs, but muſt be frequently changed. By treating this cold clay ſoil in this mode, building ſmall houſes and barns, and working with oxen, I have improved the land ſo as to be able to let it at 10s. per acre; but I muſt obſerve, that had it been titheable, the tithe alone would have utterly precluded my efforts, for the value of the tithe would often have been more than my profit.

"In making my ſubdiviſions, I divided them into ten-acre incloſures as nearly as poſſible, and the year preceding the planting quick wood, or white thorns, I prepared the ground, where the fences were intended, by frequent ploughings, and planted potatoes. In the autumn, after theſe were gathered, I made a ditch, breaſted the cam with ſtones, and planted the wood behind the cam, taking care to have the ditch on the higher ſide of the fence, ſo as to intercept the water before it reaches the roots of my quick-wood; and as warmth and ſhelter are deſirable attainments in all high expoſed ſituations, within my fences I make a border plantation about 20 feet wide, fence this off with quick-wood, and alſo fill my ſubdiviſion fences with foreſt trees moſt adapted to the ſoil. For though theſe may ultimately prove injurious both to the fences and the land, yet when that begins to be the caſe, they are eaſily taken down, and ſerve for ſtakes and bindings, when the hedges require cutting. As water is not always to be had in every ſituation where it is wanted, I make a ſquare, or round pond, where the fences interfect one another, fo as to make one pond ferve to fupply four clofes, thus :

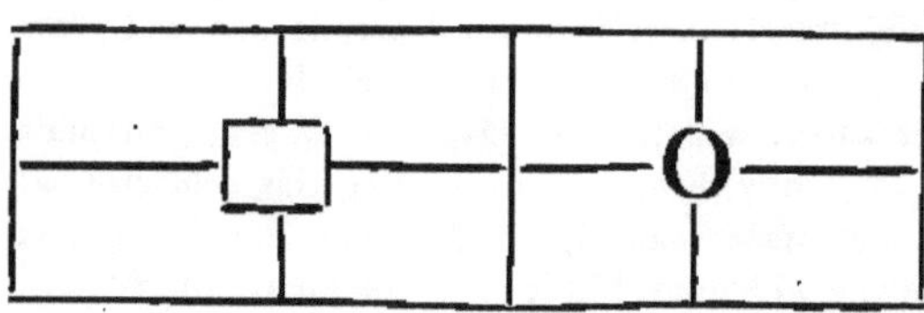

"By this means, I can either plough or pafture any clofe without inconvenience, and if the ftrata be ever fo open or porous, yet thofe ponds are eafily made retentive, by digging them deep, then laying a covering of lime, or lime afhes, at the bottom and fides, which will prevent worms and moles working; afterwards puddle it well with earth and water, and when that is got dry, pave with fmall ftones the inlets out of each clofe for the cattle to drink at; and then open ditches to let water into, and out of the ponds; and, if well executed, they will afford a due fupply of water during any dry feafon. A farm of this unkindly foil, and high fituation, will turn to beft account in having it occupied in regular courfes of one-third arable, one-third meadow, and the remainder pafture, ftocked with young breeding ftock; and by changing the land from meadow to pafture, and pafture to arable in due fucceffion, and always wintering as much or more ftock than you can fupport in fummer, you will of courfe raife confiderable portions of dung, and thereby ultimately improve the foil. This plan is, however, to be far exceeded in rapid improvement where inhabitants abound, fo as to enable you to let your property in fmall parcels, by building fmall, yet fuch buildings as are calculated to answer the purpose of any established manufactory.

"In the cultivation of my boggy allotments, I was equally erroneous in my first outset; for I rushed hastily to effect a drainage, and pursued the advice and plan of one very well versed in that operation, where the defects were only surface water, or day springs; my drains were judiciously placed, well cut, properly filled, and ample bottom apertures left; but unfortunately the nature of the springs, or water, was of the same hard incrustinating quality as the dropping well at Knaresborough, and this soon adhered to the sides, and every obstructing particle within the drains, so as to block them up; I was then obliged to open them, and suffer them to remain open, at least for a considerable time; even some of them yet emit such hard water as not to allow of covering. This occasioned much expence, and some delay; but having got the surface water off, I pared and burnt, and took rape or turnip, and a succession of oats and fallow, till I could get it into a state for grass; and then I sowed such parts as were become firm by draining, with hay seeds, and a species of clover called cow grass, being our native honeysuckle grass, which is perennial, and having a solid stem, does not contain so much fixed air as red clover, and consequently never blows cattle.

"In this state it has remained eight or ten years, is very good pasturage, and will even feed a Scotch bullock. Such parts as were too boggy to be totally corrected, I have made into willow garths, and plantations of other aquatics, which thrive tolerably well; and in a few years I have no doubt will yield considerable profit. I still keep draining them where defects appear; and when I am fully convinced the covered drains will not require opening again, and that the land will bear the operation of the plough, I will turn the ſwarth down, roll, and then ſow with oats before I harrow, afterwards harrow the ſeeds in, and roll again. The next autumn, winter fallow, and in the ſucceeding ſpring prepare the land for turnips; and in the year following, if the land is ſufficiently clean, ſow oats and hay ſeeds, cow graſs, and white clover, and then convert it to paſturage.

"When the land, which is of a looſe black earth, was laſt in turnips, it happened to be a very froſty hard winter, yet I obſerved that the turnips that grew thereon were leſs affected by the weather, and laſted good longer in the ſpring, than any that grew on much better ſoils; and this I have ſince often noticed on land of the ſame quality in other ſituations."

We have juſt one thing more to add upon this head, and that is, to ſuggeſt the propriety of declaring incloſing bills to be public acts, ſo long as the preſent ſyſtem is adhered to. We underſtand, when any legal diſpute ariſes, in conſequence of theſe bills, that the judges will receive nothing but a certifyed copy, the procuring of which is an additional expence to the parties.

NOTES on Chapter II.

(*a*) Great part of which, call loudly for improvement by the plough and the spade; may the call be obeyed, lest we *fight*, and *weave*, and *hammer*, till we have not bread to eat. *W. P.*

(*b*) Not a doubt of it; scarcely a bleak hill in the island, where wood of one kind or other would not thrive: many a spot is condemned by planters for want of ascertaining, in a small nursery' which kinds of trees will suit the soil and climate, previously to the formationof any plantation. *A Yorkshire Freeholder.*

(*c*) This a most necessary consideration, and well deserving the utmost attention of the Board; to say more on the subject than is mentioned in the text would be superfluous.

A Yorkshire Farmer.

(*d*) It is certainly true, that unstinted commons are eat up by mercenary and opulent individuals, and so overlooked, that they can be of no real service, either to themselves or others; whereas if commons were stinted, the poor cottager who could not stock his part, might receive a valuable compensation for his right. Thus a proportional stock would be put upon them, and every one receive advantage. This is only suggested, however, where inclosures cannot be accomplished; but if a general inclosure could be obtained, it would most certainly be productive of great national advantage. *A Yorkshire Farmer.*

(*e*) All the waste lands, ought to be divided as soon as possible, so as every proprietor might have an opportunity of improving his share, in one way or other. *T. H.*

(*f*) Here is a proof in point to the argument respecting small farms. *W. Fox.*

Answer—Not at all, it is only in favour of small proprietors.

R B.

CHAPTER XII.

IMPROVEMENTS.

SECT. 1.—*Draining*.

THIS most useful practice ought never to be neglected by the farmer; as, where the nature of the soil, and situation of the ground requires it, no money can be so advantageously expended (*a*). In our survey of the West Riding, we found draining was assiduously attended to, in many places; but that, in others, it was either totally neglected, or imperfectly performed: in particular, that useful measure of clearing out the water furrows, upon the tillage fields, which is absolutely necessary upon moist soils, was very negligently executed (*b*). As soon as possible after a field is either ploughed or sown, the whole furrow along the end of the field, betwixt and the head-ridge, together with such parts of the field itself where the water, from want of level, cannot get off, should be digged of a proper depth, and perfectly cleaned out. This lays the field in such a situation, that the greatest falls of rain run off immediately; and a due attention to this practice, constitutes, in a material manner, the difference betwixt the good and the bad farmer.

Hollow drains are executed in various ways. In some places the shoulder drain prevails. This is done by digging the bottom of the drain narrower than the top, and covering it with the surface sod, which may do in some cases where the sward is strong, but never can be fully depended upon. Where they are filled with stones, sometimes the largest are set upon their edge, casting inwards, till they join; which leaves a small vacuity for the water running, and they are then filled up with small stones. In other places this is done with bricks; but where plenty of materials allow it, we never could discern a more efficacious method of filling drains, than by doing it with round land stones thrown in indiscriminately, which, if care is taken that no earth is mixed amongst them, and the top well covered with straw before they are filled up, will run longer, and be less liable to interruption than when a vacuity is left by either setting the first stones upon their edge, or by walling the sides, and covering with flat stones, and at the same time is considerably cheaper.

A gentleman near Skipton writes us as follows:— "The greatest improvement I hear of is in the mode of draining, which is now done with stones above and below, and walled with them on each side: the price of this work for a yard deep, is about 1s. 6d. per rood of seven yards, including the stones, a cart load of which will complete a rood, and is worth about 3d. at the quarry. There is likewise a kind called a shoulder drain, practicable only in clay lands, which is made by using a narrower pointed spade at the bottom, which leaves a kind of shelf, or shoulder, on each side, to prevent the earth with which it is filled, from falling to the bottom: the uppermost spadeful is first laid in with the turf downwards, and then filled with the mould; the surplus (as there is always some) is either made into a compost with lime, or spread immediately upon the land. The price of this sort of draining is about 6d. per rood, at a yard deep; and so on in proportion. The drains, before these abovementioned were introduced, were usually covered with brush-wood, or perhaps straw or rushes."

Sect. 2.—*Irrigation.*

In many parts, especially in the manufacturing district, great improvement is made upon the grass fields, by watering or floating them (c). Mr Walker, at Crow-nest, is the most particular in this respect, and has his water so admirably disposed, that he can float the greatest part of his fields, whenever he thinks convenient. We do not pretend to be acquainted with this branch of husbandry; but in some places, we were told, its advantages were equal to a top-dressing of manure.

Mr Ellershaw, at Chaple-le-dale, near Ingleton, gave us a particular account of the manner used by him, and several of his neighbours, to water their fields. They float it early in the spring, which rots the moss, enriches the ground, and consequently produces an additional quantity of grass. Where a sufficient quantity of water can be got, and proper levels found, it certainly is the cheapest and probably the most efficacious way of enriching ground. After all, a good deal of judgment is required to perform this operation in a proper manner.

Sect. 3.—*Paring and Burning.*

Our information on this head, was various and contradictory. In some places, the practice is prohibited, unless with the consent of the proprietor. In others, it is deemed the best method for breaking up all grass grounds, and is not supposed to waste the soil in any shape. Our opinion is, that upon some grounds, paring and burning may be good management, particularly upon rough coarse sward, which cannot otherwise be easily brought into a proper state of cultivation. But that upon the whole, it is a practice that should be cautiously used, as it tends in a material degree to exhaust and impoverish the soil (*d*). The expence of paring and burning of land, with the spreading of the ashes, is from 18s. to 24s. per acre.

An intelligent gentleman, in the neighbourhood of Doncaster, has favoured us with his sentiments on the most proper soils for paring and burning.

" All old grass fields, which are husky and will not easily pulverise, ought to be pared and burned, as no land is proper for raising corn before it is thoroughly reduced; and this is accomplished in a more speedy manner by paring and burning, than by any other process. The soil is also enriched by these operations, grubs, and worms are destroyed, and a fermentation occasioned, something similar to that of yeast amongst flour and water. Grubs and worms prevail much in all these old grass fields; and, before they are extirpated, the corn sown upon them is in danger of perishing, for want of proper food. This is a fact well known to every practical farmer who has broke up such soils.

" *Limestone and Heath lands* are well adapted for paring and burning, as they are generally poor; and the ashes, by acting as a manure, produce good crops afterwards.

" *Carr-land, or peat earth*, suit for paring and burning best of all, as it is difficult to pulverise such soils in any other way. The roots of the herbage which grows upon them, are so strong that the surface is thereby bound fast like a matt, but paring and burning removes this obstruction at once, and consequently ought to be adopted when soils of this kind are brought into cultivation.

"*Sand soils* less suite for paring and burning, as sand is an expeller of fire, and will not burn to ashes. Clay is also improper for burning, for it thereby becomes brick; nor can I recommend the hazel earths, which generally carry a fine swarth, as being suitable for paring and burning; they are not difficult to pulverise, and are soon brought into a proper state for carrying corn crops."

Sect. 4.—*Manures.*

This is a subject which deserves particular attention as it is upon the solid foundation of manuring that every good system of husbandry must be built.

The manures used in the West Riding, besides those generally used in other parts of the kingdom, are bones, horn shavings, and rape dust, with several other articles of refuse from the manufacturing towns; and from the accounts we received, the effects of these extraordinary manures are highly beneficial. With regard to the lime husbandry, and the collection and application of home-made dung, we apprehend the practice of the district is very faulty, and we shall give our reasons for this opinion.

1st, In the pasture parts of the country, the hay is consumed upon the field, and from its being thrown indiscriminately upon the ground, the dung may be said to be in great measure lost (*r*), at least the value of it is much reduced in comparison to what it would be, if the hay was eaten at home in the house, or the yard; and the dung carefully collected together in a heap, so as fermentation might properly take place. We decidedly condemn the eating hay in the field, as occasioning great waste of that necessary article, independent of the loss sustained by the improper application of the dung.

2dly, The home-made dung, in the above parts of the country, is generally laid upon the rich pasture fields, which have been cut that season for hay, and not upon the tillage lands. We have doubts, whether dung can ever be applied with equal propriety, as upon well wrought fallows. If the dung exceeds the quantity necessary for the fallows, which in few situations will be the case, it ought to be laid upon other parts of the farm, which are under the plough, and not upon the grass fields, which when properly sown down, will sufficiently improve themselves (*f*).

In the corn districts, dung is applied with more judgment, it being generally laid upon the fallow or turnip break, though even there it is sometimes laid upon the grass. We are of opinion, a great deal more dung might be accumulated, if the stubbles were cut lower than is presently done (*g*). Barley and oats are often cut with the scythe, which so far obviates this argument; but wheat, which is the prevailing crop, is always cut with the sickle.

From not seeing the crops upon the ground we cannot say with precision what proportion of the straw might be left. But, from a careful examination of the stubbles, we suppose it at least to be one-third (*h*). This not only occasions a great loss of grain, as all the straggling heads are thereby left, but also deprives the farmer of a large portion of home manure, for the dry stubble, left upon the field, will never ferment; it is therefore of no use to enrich the ground, and occasions great inconvenience, when the land is ploughed down afterwards.

To ascertain the difference betwixt high and low cutting, an experiment was made upon part of a field of wheat, two ridges of which, were cut close by the ground, and the other two confiderably higher, though not fo high as the general run of the Yorkfhire ftubbles. Each of the divifions was apparently of equal quality, and meafured a trifle more than a quarter of a Scotch acre, which is above one fifth larger than the Englifh ftatute acre. The crop was ftooked feparately, and the time taken to the part cut low, was 1 hour and 24 minutes, of 8 fhearers, while the high cutting was performed by the fame number of hands, in 48 minutes. The wages paid that week were 18d. per day, and the fuppofed expence of maintenance 6d. or 2s. per day altogether. When threfhed, the grain and ftraw were carefully meafured and weighed, and the refult of the experiment was as follows:

Refult—8 fhearers, 1 hour 24 minutes, at 2s. per day, or 22½d. per hour -	L. 0	2	4
The fame hands 48 minutes - -	0	1	4
Difference of expence - - -	0	1	0

in favour of high cutting one fhilling, or four fhillings per acre.

1¼ pecks of wheat more upon the low cutted ridges, than thofe cutted high at 1s. 4d. per peck - - -	L. 0	1	8
14 Stones (22 averdupois pounds) of more ftraw, at 2d. per ftone - -	0	2	4
	0	4	0

or fixteen fhilings per Scotch acre.

From which deduct the increafed expence of cutting, there remains a benefit of twelve fhillings per acre in favour of low cutting.

The above trial according to the beft of our judgment,

was fairly made, and the reason which urged us to make it, was to silence the objections of some neighbours, who alledged low shearing was not profitable. It is proper to observe, that the field of wheat on which the trial was made, was not broke down nor straggled, so was in a favourable condition for high cutting. We have seen wheat fields, where three times the quantity might have been left, unless great pains were used. Barley is another grain that requires careful handling, as, where the bottom is rough, or the straw short, it is almost impossible to make good work. The utility of taking care of a crop is so evident, that we presume it is unnecessary to urge arguments in favour of what few will contradict, although they have not patience to practise it. What can be more absurd, than for a farmer to carry on all the previous operations with accuracy, and when the object of his labour is come to perfection, to allow it to be hashed and mangled by his shearers at harvest?

The farmer is in many cases deprived of a due quntity of dung, by keeping too many cattle. We venture to lay it down as a rule, that no greater number should be kept, than is necessary to reduce the straw to putrefaction. When more are kept, although the quality of the dung may be improved, yet the quantity is curtailed.

Bone dust, or as it is called, *hand tillage*, is used to great extent upon all the fields for twenty miles round Sheffield. Bones of all kinds are gathered with the greatest industry, and are even imported from distant places. They are broke through a mill made for that purpose; are sometimes laid on the ground without any mixture; but it is supposed most advantageous to mix them up with rich earth, into a compost, and when fermentation has taken place, is the proper time to lay them on the ground. We also heard of another manure, which can never be more than a local one, viz the refuse of hogs bristles from the brush manufactories. One gentleman informed us that he had manured four acres with this refuse, and that its effect greatly surpassed that of street dung, which the rest of the field had been covered with.

Lime is applied to the greatest part of the land in cultivation, and the quantity laid on at one time, is so inconsiderable, that in our humble opinion, it can never produce the intended effect. Whenever we speak against a general practice, we do it with diffidence; but upon this occasion, we cannot refrain from expressing our dissatisfaction, both with the quantity applied, and the frequent repetition of this article (*i*).

The farmer is too often obliged, by the covenants subsisting between him and his landlord, to throw lime upon land, where, in the real sense of the word, it is thrown away. It must appear exceedingly absurd to any person, who knows the manner in which lime operates, and the number of years its effects continue, that the farmer should be obliged to lime his land every third year, whether it needs it or not. This is done by every lease, where two crops are only allowed to a fallow, and where it is covenanted to lay lime upon that fallow. The specified quantity is in many cases so small, being sometimes one chalder, or a chalder and a half, that it never can produce effects adequate to the expence, or indeed any expence at all. It may be said, that by frequently laying on small quantities, that a sufficient dose is given at last. This argument is plausible, but it should be remembered, that the effects of the first partial liming, is probably wore off, before the second comes to its assistance; and that if the first is stimulating and fermenting the land, the second is only a prodigal waste of expence.

But why oblige the farmer to lay lime upon his land at all? If it be for his intereft, he will do it without any obligatory claufe in his leafe; if it is not for his intereft, a burden is laid on his fhoulders, that can give benefit to none. It is furprifing, proprietors fhould infift upon this; for lime has never been underftood to improve the *real* value of the foil, in a permanent manner, but is generally confidered as a ftimulus, or ufed to procure a temporary exertion.

We were particularly anxious to afcertain the quantity of lime laid upon an acre, and we found it to be, in different places, from 1 chalder, or 32 bufhels, to 100 bufhels. Some people may ufe rather more, but from 60 to 70 bufhels per acre, may be regarded as an average; a quantity very inadequate, in our humble opinion, to the intended purpofe.

Lime, in the Weft Riding, is principally applied to fallow, and fpread upon the ground immediately before the laft ploughing. We judge, unlefs in fome particular cafes, it would be ufed with as much advantage and with greater convenience upon the grafs fields. For inftance, inftead of laying it upon the fallow, preparatory to turnips, or upon the clean fummer fallow, let it be laid upon the clover crop, which is the third of the ufual fequence; or, upon the pafture lands, previous to breaking them up for corn (*k*). The land is generally at that time in a fituation proper for the operation of lime, and it can be applied, at different periods, with lefs trouble and inconvenience to the farmer.

In no practice whatever, has greater errors been committed, than in the management of land, after it has been limed. This manure as it is called by fome perfons, or ftimulus as it is called by others, has been ufed with wonderful fuccefs, in every part of the ifland,

and was known to the ancient Britons before the landing of the Romans under Julius Cæsar. From its effects, the strongest soils are rendered free and pliable, while those of an opposite quality, are rendered compact and firm. Although the consequences attending lime have generally proved very beneficial, in the first instance, (lime being often as superior to dung, as dung is to nothing,) yet great errors have been committed in the after mode of management, by persisting in corn crops till the land was exhausted. When land is reduced to this state, it will be found just as much lost money to give it a second dose, before it is enriched by dung, or refreshed with grass, when a repetition may be given with certain advantage.

The quantity of lime we have been in use of laying on an acre of ground, is from 250 to 300 bushels, which is from 200 to 240 bushels to the English statute acre; but much depends upon the nature of the soil, upon its present condition, upon the quality of the lime, and upon its being properly incorporated with the soil. It is at all times safer to increase than to diminish the above quantities, as an over-dose is seldom hurtful, especially upon strong soils.

Much dispute has taken place upon the best mode of applying lime, whether in a hot powdered state, or when it is *effete*? whether on grass or fallow land? and, when applied to the former, whether this ought to be done a year or two before it is broke up? The writer of this report, who probably has limed as much ground as any of his profession, can with confidence maintain, that where the land is in that state, which constitutionally disposes it to receive benefit from the application of calcareous matter, *that is* when it has lain long in grass or is sufficiently enriched with dung, or other manures, that lime will in these cases operate in whatever form, or upon whatever surface it is applied. It is certainly the thriftiest way of using it, to lay it on in its powdered state, and probably it gets sooner into action, when administered to a well pulverised fallow; but that the consequences will be equally beneficial in a year or two, that is as soon as the lime is fairly mixed with the soil, we have not the smallest doubt.

It has been thought that when lime is applied to grass lands, the safest way of doing it, is to lay it on a year or two before the field is intended to be broke up, otherwise the lime will be buried in the bottom of the furrow. The writer once tried an experiment to ascertain the fact. He limed thirty acres at the rate of 250 bushels per acre. A part was limed three years before ploughing, another part two years, another part one year, and the remainder about 8 or 10 days before the ploughing commenced. At harvest the whole crop which was oats was equally good, and the best proof that the lime had commenced operation was, that twenty acres of the same field not limed, were full 15 bushels per acre short of the quantity produced on the rest of the field.

We have had occasion to see that lime is useless upon some soils, chiefly those of a moorish or soft nature, which have been previously limed and hard cropped afterwards; but if a similar quantity was wrought up in a compost of earth, &c. that the consequences were highly beneficial. We are inclined to think, this is the safest way of repeating lime upon soils that do not possess much vegetable substances; at least, from trials repeatedly made, we have never been disappointed.

If these composts are made up on the head ridge of the field, or on any rich land adjoining, and wrought wholly with the plough and harrow, they are not more expensive than ordinary manures. The great object is to save carriage, as from the quantity required to cover an acre, the charge is confiderable when brought from any diftance.

It is believed that theoretical writers are often much miftaken with regard to the nature and operation of lime. Indeed few branches of agricultural fcience are lefs underftood, and we may venture to fay the fubject will not be better underftood, without reforting to a body of *facts*. Many of our writers are like the philofophers who figured before Lord Bacon's day; they form a theory, and bring their facts to that ftandard, inftead of building their theories upon the folid foundation of facts and experiments.

Judging upon thefe principles, that more ufeful information will be communicated to the public, by a practical paper upon this important fubject, than by a pompous parade of philofophical knowledge couched in technical terms; we therefore give a place to the following paper, upon the application of lime, furnifhed us by a farmer in this country, who has ufed lime to a great extent, and attentively marked the progrefs of its operation upon a variety of foils.

"In the year 1778, I limed a field, the foil of which was principally compofed of thin clay, upon a bottom retentive of moifture. The field was fallowed from grafs, and the lime which was completely *effete*, or wet, was applied in the fpring thereafter, at the rate of 45 bolls per acre. The field was fown with oats, but no benefit was received the firft year from the application. The next year the ground turned loofer, and a ftrong fermentation took place, the effects of which have not yet entirely ceafed.

"The fame year I limed a field of real moorifh foil, which had formerly been over-cropped, after the application of lime. The land was fummer fallowed, and the lime laid on the next fpring, when it was *effete*, but instead of producing beneficial consequences, the crops have repeatedly *failed*, and the value of the lime may be considered as lost.

"1780. Limed another field of the same quality, but the lime was applied hot. The same consequences followed, as in the field last mentioned. After being fallowed and dunged, the lime appeared to operate, but not to a sufficient extent for defraying the expence.

"1781. Fallowed a field of moorish soil, which had formerly been limed; tried lime upon a part of it, which was laid on hot; very little difference however appeared betwixt the crops of corn and grass upon either parts; the same year limed a field of old grass upon the surface, which carried no marks of having ever been limed. The soil was partly a thin clay, and partly of a soft sandy nature, but all incumbent upon a wet bottom. The effects were trifling for the first and second years, but after being completely fallowed, the consequences were astonishing. It has since that time been twice in grass, and the lime continues in full vigour of action.

"1784. Limed part of a field of a soft loamy soil, upon a wet bottom, when under summer fallow. The lime was laid on dry, and operated the first year. In some seasons the crop upon the limed part, has been nearly double more bulk than upon the unlimed part.

"1787. Limed a fallow field which had been lately in old pasture, composed of strong loam incumbent upon clay. The lime was laid on hot before harvest, and appeared to operate immediately. The crops have been uniformly good since that time, although it has been in grass but one year. The same year, covered a field of the same soil with a compost of lime and earth, which produced effects not inferior to those upon the last mentioned field. The quantity of lime used was 20 bolls per acre, which was spread upon a high broad headridge, frequently turned by the plough, and laid upon the ground after being summer fallowed.

"1789. Limed a field of grass land composed of thin sharp loam; the lime was laid on hot before winter, and its effects appeared upon the first crop.

"1790. Limed a considerable part of a large field that had been four years in grass; the soil principally loam, but of several varieties, and the lime was laid on at different times, but the whole operated equally well the first crop. The succeeding year, what had not been limed was summer fallowed; the half of which was then limed, which has answered equally well, while the crops upon the part unlimed are greatly inferior.

"1791. Limed a grass field of soft loam, which was ploughed the following year; the lime was *effete* when applied, and operated immediately.

"1794. Limed a grass field of thin clay; the lime was completely *effete*, and promises to answer well.

"1795. Limed another grass field of much the same soil; the greatest part of the lime was laid on in a hot powdered state, but the remainder was *effete*. From the fermentation which has taken place over the whole field, it appears to operate equally well, in whatever state it was administered.

"From the above account the following inferences are drawn:

"*1st*, That the application of lime, to moorish soils which have been already limed, is an unprofitable business.

"*2dly*, That where the constitution of the ground is disposed to receive benefit from lime, it may be applied either hot or *effete*; upon grass land or upon fallow.

"*3dly*, That lime is equally beneficial to all sort of soils, provided they are in a proper condition for receiving the application."

N. B. It is the Scots acre that is always meant in the above paper, 4 of which are equal to 5 English statute acres, and the size of the *boll* mentioned is nearly equal to six Winchester bushels.

A farmer in the West Riding whose opinion we highly respect, writes us upon this subject in the following words:

"Lime husbandry was more practised some time past than at present; for it is found, that where lands have been long under the plough, and often dressed over with it (which has been the general practice for a century past), it has very little effect. The old farmers used no other tillage, till very lately, but what was made in the farm-yard, and many of them no other yet, always liming their clay land fallows, and sowing wheat; next oats, beans, or broad clover, and again wheat. They have thus fallowed and limed, again and again, for 30 or 40 years together, laying on at the rate of about 120 bushels of Knottingly stone-lime upon an acre, which will be two four-horse cart loads. This stone is brought from near Pontefract, about 15 miles by water. Since we got the navigation, it is burnt by the river side, about 3 miles distance from us: it costs at the kiln about 4½d. per bushel; the expence of conveyance from the kiln to the land (to average a circuit of six miles) will be about 1d. per bushel, and the expence of watering and spreading nearly ½d.: so that the whole expence will be about 6d. per bushel, or L. 3 for a statute acre. This is collected during the summer, and spread on at any convenient time, a little before wheat sowing.

"But, in my opinion, this time is too late, as I find the sooner it is spread on in the spring, and the oftener it is ploughed afterwards, the more intimately it gets mixed with the earth; having perfectly absorbed its own air and water, the better it fertilizes the foil, and fits it for the produce of a crop. The feafon of laying it on is not however regarded by the generality of farmers, nor fcarcely any other property refpecting it, but convenience for their other employments. The moft improved method I am acquainted with, and which I find to anfwer beft, is to lay upon clay foils about 180 or 200 bufhels of Knottingly ftone-lime upon an acre. This ftone, upon being analyzed, is found to be mixed with a ftrong fand, about one-third of its weight (for we have two forts of lime of very different properties). The earlier in the fummer it is laid on, the better, for the fallow to receive a few ploughings afterwards. It alfo anfwers beft to be laid on the firft fallow after feeds, as the frefher the land, the greater its effects. I think it not prudent to lime two fallows together, except there has been an interval of reft, and other manures fpread on in the mean time; nor do I find it anfwer upon old ploughed wore out foils. Hence arifes the philofophical opinion of fome ingenious farmers, that lime, poffeffing neither oils nor falts, acts only as a ftimulus or forcer to other manures, bringing fuch vegetative qualities, as are in the foil, into more powerful life and activity. Upon dry land that is proper for turnips, I lay 80 or 100 bufhels of Emfall lime per acre. This is mixed with a ftrong clay about the fame proportion, as the other of fand; there is fome cauftic quality mixed with this lime, that if too great a quantity be laid on, inftead of affifting it, deftroys vegetation: but about this quantity is helpful, it ftiffens the ftraw, makes it ftand firmer at the root, and heavier in the ear. I do not ufe this as a complete, but only an affiftant dreffing betwixt fallows; laying it on in the autumn before the laft crop before fallow, as foon as poffible after the preceding crop is reaped. I then plow down and fow with either wheat or oats, to either of which it is helpful, and the following year will be more serviceable to the turnip crop, than if spread on the land the same summer. This lime costs about the same price as the other. It is to be observed, that these lands are kept altogether fresh by being sown with seeds, and pastured with sheep every other fallow; and always dressed with bones or fold manure, or both, for turnips."

Mr Peach at Sheffield informed us, that the lime brought from the neighbourhood of Doncaster, would not answer upon his land; but that 80 or 90 bushels per acre of the Derbyshire lime operated well. This confirms what we have already said relative to the theory of lime being imperfectly understood. Indeed the liming of land being an expensive business; where quantities such as from 2 to 300 bushels are laid on an acre, every person should previously ascertain the qualities of the lime and consider attentively the nature of the soil upon which the application is to be made.

Sect. 6.—*Warping of Land.*

This is a mode of improvement which produces the most beneficial effects, and originated, we believe, in the district under consideration. It is obvious the practice must always be a local one, for it is only in a very few situations where it can be adopted, but wherever circumstances permit it to be practised, we cannot recommend such a measure in too strong terms. The fact is that a soil of the richest quality may thereby be created, which may be made of any depth thought necessary, and the poorest and most barren soils may be rendered as fertile and productive as those of a different description, without a halfpenny of more expence being incurred in the one cafe than in the other.

Upon this important fubject we have received three very valuable communications. The firft is tranfmitted by the Right Honourable Lord Hawke, who has conftantly difplayed the greateft zeal to render this work as complete as poffible. The fecond is from Mr Day at Doncafter. And the third from a worthy friend to whom we have, upon many occafions, been under the greateft obligations.

Obfervations on Warping Land, tranfmitted by the Right Honourable Lord Hawke.

" The land to be warped muft be banked round againft the river. The banks are made of the earth taken on the fpot from the land: they muft flope fix feet; that is three feet on each fide of their top or crown of the bank, for every foot perpendicular of rife: Their top or crown is broader or narrower, according to the impetuofity of the tide, and the weight and quantity of water; and it extends from two feet to twelve: Their height is regulated by the height, to which the fpring tides flow, fo as to exclude or let them in at pleafure. In thofe banks, there are more or fewer openings, according to the fize of the ground to be warped, and to the choife of the occupier, but in general they have only two fluices, one called the flood gate to admit, the other called the clough to let off, the water gently; thefe are enough for ten or fifteen acres: When the fpring tide begins to ebb, the flood gate is opened to admit the tide, the clough having been previoufly fhut by the weight of water brought up the river by the flow of the tide. As the tide ebbs down the river, the weight or preffure of water being taken from the outfide of the clough next the river, the tide water that has been previously admitted by the flood gate opens the clough again, and discharges itself slowly but completely through it. The cloughs are so constructed as to let the water run off, between the ebb of the tide admitted, and the flow of the next; and to this point particular attention is paid: The flood gates are placed so high as only to let in the spring tides when opened. They are placed above the level of the common tides.

" Willows are also occasionally planted on the front of the banks to break the force of the tide, and defend the banks by raising the front of them with warp thus collected and accumulated: But these willows must never be planted on the banks, as they would destroy the banks by giving the winds power to shake them.

" The land warped is of every quality; but to be properly warped it must be situated within the reach of the spring tides, and on a level lower than the level of their flow. The land in general is not warped above one year in seven, a year's warping will do for that time.

" The land is as other land, various as to the preference of grain to be sown on it.

" Land has been raised considerably by warping: One field of bad corn-land, good for nothing, was raised in three years fourteen inches: It lay idle for that time that it might be raised by warping, it was sown with beans last year, and promised by appearance a crop of eight quarters. If possible this shall be ascertained as to the quantity threshed.

" The warp consists of the mud and salts deposited by the ebbing tide: Near Howden one tide will deposite an inch of mud, and this deposite is more or less according to the distance of the place from the Humber.

" Cherry Cob sands were gained from the Humber by warping: They are supposed to be four yards thick of warp at least: Some of those were ploughed for twelve, fourteen, or sixteen years, before they would grow grass seeds: The greater part is now in feeding land, and makes very fine pastures.

" The land must be in tillage for some considerable time after warping, for six years at least: The land if laid down to grass, and continued in grass, is not warped; for the salts in the mud would infallibly kill the grass seeds.

" When it is proposed to sow the land again with corn, then the land is warped: When they find the grass decline, then they warp and plough it out: As the land varies in quality, so does the time during which it will produce good grass: The land is never fallowed but in the year when it is warped.

" For a view of a clough see Mr Young's Northern Tour, first vol. plate 3. p. 212. The flood gates and sluices for letting in the water are like the common sluices and gates in canals for raising the water to assist the passage of boats; sometimes also the flood gates or sluices are placed above the clough perpendicularly."

Information from Mr Day of Doncaster concerning the Warping of land.

" The practice of warping, in the low part of the West Riding of Yorkshire, I conceive, originated from the tides overflowing the banks of the rivers, and thereby leaving a sediment, which was found to be excellent manure and that the land brought very large crops after being flooded in that manner, Indeed I believe the first trial of warping was made by a small farmer, who had some low land adjoining a certain river called the Dutch river, which was very poor soil, the lowest part of which was levelled with the highest, by the overflowing of some very high tides, which convinced the farmer that he could, by banking the land round, and laying a tunnel through the bank of the river, raise the same, and make it of considerable more value. He therefore applied to the commissioners of sewers for the level of Hatfield chase, (being commissioners appointed for draining that part of the country &c.) to grant him an order giving him leave to lay a tunnel, a few inches square, through the bank of the said river, for the purpose of warping his land, which was granted him (with a great deal of reluctance, for fear of overflowing the country with water) on his giving a proper security for indemnifying the country against any injury which might happen thereby, which answered his purpose extremely well. But now there are cloughs laid of 6 or 8 feet wide, and drains made of proper dimensions, to convey the water accordingly. I am not certain how long it is since warping came much into practice; but however it is not many years ago; I believe not more than 20 or 25 years or thereabouts.

"As to the expence of warping, it is an impossibility to make any estimate without viewing the situation of the lands to be warped, and the course and distance it will be necessary to carry the warp to such lands, as, *1st*, The situation of the lands must be considered. *2dly*, The quantity of land the same drains and cloughs will be sufficient to warp. *3dly*, The expence of building the cloughs, cutting the drains, embanking the lands &c. An estimate of which expence being made, then it will be necessary to know the number of acres such cloughs and drains will warp, before any estimate per acre can be made; therefore you will easily conceive the greater quantity of land, the same cloughs and drains will warp, the easier the expence will be per acre. In my opinion there are great quantities of land in the country, which might be warped at so small an expenc, as from L. 4 to L. 8 per acre, which is nothing in comparison to the advantages which arise from it. I have known land which has been raised in value by warping, from L. 5 to upwards of L. 40 and L. 50 per acre; therefore it is easy to conceive that the greatest advantages arise upon the worst land, and the more porous the soil the better, as the wet filters through, and sooner becomes fit for use.

"The advantages of warping are very great; as, after lands have been properly warped, they are so enriched thereby, that they will bring very large crops for several years afterwards, without any manure; and, when it is necessary, the lands might be warped again, by opening the old drains, which would be done at a very trifling expence, and would bring crops in succession for many years, with very little or no tillage at all, if the lands were kept free from quick grass, and other weeds, which which must be the case in all lands where they are properly managed; besides the drains which are made for the purpose of warping, are the best drains that can be constructed for draining the lands at the time they are not used for warping, which is another very great advantage in low lands.

"As to the disadvantages in warping, I conceive there can be very few, if any, as the land might be warped in the year that it ought to be a summer fallow. Indeed all lands that are warped, ought to be prepared in the spring as fallow lands, so that they are ready to let in the warp by the month of June, as the three succeeding months, are the most proper months in the year for warping, (but they might continue warping longer when it is necessary, therefore the rent is out of the question. The only inconveniences that can arise, in my opinion, are from the blowing up of the cloughs, or breaking of the banks, (which is seldom the case but where there is some neglect in the works,) and thereby overflowing the adjoining lands, and very probably deſtroying the crops; but it nevertheleſs very much enriches the land that it overflows; however, theſe circumſtances ſhould be guarded againſt by every cautious engineer.

"Warped land ſeldom fails of carrying good crops; but oats are moſt to be depended on the firſt ſeaſon. I think warped land is better calculated to grow oats, wheat, and beans, than barley, as the ſoil by that means is ſo very rich, that barley generally grows too coarſe. It never fails growing artificial ſeeds of all kinds, and is the beſt of paſture land.

"Land once well warped will laſt a number of years; but in my opinion where conveniency ſerves, the beſt way is to lay on a little warp every time it becomes fallow, which if kept in arable land, would be about every 5 or 6 years, and by that means the farmer would ſeldom fail of having great crops. In ſhort I know no ſort of management ſo cheap as warping, when properly applied."

Mr Day of Doncaſter's anſwers to the queries on his former obſervations on Warping Land.

Anſwer to Query 1ſt. Warp, is the ſedement left upon the land by flooding the ſame with tide water. Letting in the water is alſo called warping, from the ſediment which the water leaves behind it, which is called warp. Letting in freſh water, not being tide water, would not be called warping, but flooding the land.

Anſ. to Qu. 2d. The water, being tide water, and coming from the ſea or large rives, is of courſe brackiſh, and the warp or ſediment it depoſits is of the ſame nature. Freſh water, though very uſeful upon ſome land, at proper ſeaſons of the year, would by no means anſwer the ſame purpoſe as water coming from the rivers where the tide flows, as it never could depoſit a ſufficient fediment, neither would it be of half fo rich a nature as what is left by tide water.

Anf. to Qu. 3d. The water does not at all ly flagnate, nor is it unwholfome to the neighbourhood, as it goes off and returns regularly every tide; it only continues a little time, till the greateft part of the fediment has fubfided, and then returns through the fame drain, clough, or fluice, it came from; or, if convenient, through fome other fluice or inlet made for that purpofe.

Anf. to Qu. 4th. The drains are open drains, and cut the fame as all other drains, for the purpofe of draining lands. The depth of the drain is according to the level of the land, with the river from which you take your warp; and the width agreeable to the quantity of land you mean to warp at one time, and the clough or fluice which communicates with the river.

Anf. to Qu. 5th. June, July, and Auguft, are thought the beft months for warping, on account of their generally being the dryeft months in the year; they might warp land in any month in the year, when the feafon is dry, and the frefh water in the river very low. But, if the feafon is wet, and the rivers full of frefh water, it mixes with the tide, and makes it not half fo thick and muddy, and of courfe hinders it from leaving one half or one fourth the fediment upon the land, it would in a dry feafon of the year; neither is the water got fo readily off the land in wet feafons as dry. Warping land in the fpring, can anfwer no better purpofe than fummer, as there could be no crop that year, for the warp muft ly to foak and dry, before the land can be cultivated to any advantage.

Anf. to Qu. 6th. Warped land is fuppofed to be the beft of land for potatoes, and the moft productive.

Anf. to Qu. 7th. The depth of the water upon the

land, entirely depends upon the level of the land, and the height of the tide in the river, from whence the water is taken; but, where it can be accomplished, it might be 3 or 4 foot deep or upwards, as the deeper the water, the more sediment is left; but land may be warped with a deal of less water, as it is only letting on more tides, and taking longer time to the work; it does not at all signify whether the water is always kept at the same height or not, only take care that it does not overflow the banks.

Anf. to Qu. 8th. Mr Richard Jennings of Armin, near Howden, was the first person who tried the experiment of warping, about 50 years ago. It was next attempted by a Mr Farham, steward to ——— Twisleton, Esq; of Rawcliffe, also by a Mr Mould of Potter Grange, both about 40 years ago; and it has been tried by a great variety of people since that time, to their great advantage.

Anf. to Qu. 9th. What is meant by *warping being found to be excellent tillage?* is no more than that it is excellent manure, and good for all kinds of land where it can be accomplished.

Anf. to Qu. 10th. Cloughs, what are they? A clough is an inlet cut in the bank of the river, walled on each side with a strong wall and floodgate fixed in the middle, for the purpose of letting in and out the water, and is commonly called a clough or sluice; it is nearly upon the same principle as what are used at water mills.

Observations upon Warping transmitted by a West Riding Farmer.

"Low land, capable of being flooded by the rising and falling of river tides, is of all others the most improveable. The ground is thereby enriched; no person is injured; and the benefit received is lasting and durable.

"This improvement is performed by having a sluice in

the banks to let in the water when the tide is up, and to carry it off again at pleasure, when the sediment of the water is deposited upon the surface. When this improvement is intended, the ground must be first banked up, and the cutts necessary for carrying off the water should be so constructed as to make partition fences. It is of no consequence what the soil is before it is warped, as the warp is raised as deep as you think fit, or that is necessary for growing crops. The best potatoe soil, both as to quantity and quality, is thereby produced, and it answers equally well for all kinds of grain.

" I shall now say a few words upon another branch of what may be called the same subject, viz. the great losses sustained upon the fertile fields lying contiguous to rivers by floods, at different seasons, particularly when a storm of frost and snow breaks up, which in many cases might be prevented at little expence. The fault proceeds from the want of a speedy outlet to the general receiver, and from not having cutts or drains alongst the foot or bottom of the high grounds, for intercepting the torrents which then issue from the hills. If these cutts were made in proper situations, the superfluous water would be prevented from spreading over the low grounds; thousands of acres of fine corn would be preserved to the community; the farmer, saved in many instances, from ruin, and the interest of the proprietor much promoted; for it is demonstrable, that lands in such hazardous situations, are unable to pay the rent they are capable of doing, if preserved from these destructive devastations. Another advantage from these cutts would be, that the farmer would be enabled to water his lands at the proper season, which would be highly beneficial to him; but before this can be done with propriety, the land ought previously to be laid dry; otherwise the full advantages of irrigation will not be procured. I don't think ploughed land ought to be watered, as it deſtroys the crop, beggars the occupier, and robs the dunghill. Whereas, when water is meant to improve, it ought to be kept running in a gradual way over the field, to the deepneſs of two inches, and not allowed to remain ſtationary.

"I may add, that if all the loſſes ſuſtained by the floods I have mentioned, were added together, the expences of the cutts recommended would ſoon be balanced. I have known inſtances of L. 20 to L. 100 worth of manure ſwept away at once, beſides the great quantity of ſoil carried away, which will not admit of a calculation."

We cannot finiſh this ſection, without recommending, in the moſt earneſt manner, the practice of warping, where circumſtances will allow it. It is, without diſpute, an improvement of the firſt importance: It is accompliſhed at a leſs expence than what manure, in any ſituation, can be purchaſed. By it, in fact, a new ſoil is created, and that of a quality ſuperior to that of the moſt valuable ſoils. We truſt the information here communicated, will contribute to facilitate its introduction into other diſtricts of the kingdom.

NOTES on Chap. 12.

(*a*) The draining of tillage lands must be essentially necessary, but I doubt if it is of any advantage in old pastures, as, in a dry summer, those parts which are springy, are obviously of most service in the support of cattle; where sheep are kept as a breeding flock, such places may prove pernicious by causing a rot. A little wet, and a little dry land is certainly very useful; if I could float my land with water when I plowed, I would have all springs taken off. *A Yorkshire Farmer.*

(*b*) Short leases are often the pretence, though they are seldom if ever the cause of bad husbandry. He who will not, when *a tenant at will*, carry the water off the furrows of his corn field, would not be a good farmer, if he had a lease of 100 years. *Anonymous.*

Answer.—Short leases are not assigned in the text, as the cause of draining being neglected, therefore, the censure bestowed upon the farmer by the above note, is evidently misplaced. *R. B.*

(*c*) This must be a most excellent expedient against a dry summer; about 35 years ago, I knew a few acres of land, over which waters had been forced, prove the chief support of 40 or 50 cows, during the whole summer, which was remarkably dry; and it is certain that meadows, under this management, will, upon an average, produce more hay by a ton per acre, than other lands not under this mode of management, though of equal quality. *A Yorkshire Farmer.*

(*d*) Paring an old sward which has lain for time immemorial can never waste the soil. I have a sod in my house, which I have kept for some years as a proof of that; it was not less than 4 inches thick, when first pared, of entire roots or turf, and from no inferior soil. The land from which it was taken, might have been pared twice over, and well burnt, without lessening the soil at all, and no doubt the land must be greatly enriched by the ashes produced from such a thickness of turf. On lands which have not lain a sufficient time, to produce roots for a sod, I disapprove of paring and burning. *A Yorkshire Farmer.*

(*e*) This seems to be a faulty practice. *T. York Esq.*

(*f*) Few meadows, even of the richest quality, can be found, which would not be utterly impoverished by this management.

Turnip fallows require dung, excepting on very rich and fresh land, but see the survey of the North Riding p. 33d. Potatoes likewise dung; in either case, barley properly follows. Wheat, oats, and beans, may be grown succesfsfully by means of well worked fallows, and the assistance of lime, marl, and several other manures, provided the land was not exhausted by too long continuance in tillage, and that part of the dung, which hath not been applied to the turnip or potatoe fallows, may properly be reserved in order to restore, in some degree, the riches of the meadows, which have been taken from them by mowing; in short, the whole farm may be greatly enriched by a judicious arrangement of crops, and by seasonable and successive reliefs of its arable parts. *T. York Esq.*

Answer.—The two first lines of the above observation, only fall to be considered; the remainder being void of the question. By meadows the writer must mean those fields of pasture cut for hay. This mode of raising hay, the surveyors detest, and it ought not to have escaped the attention of the writer, that according to the system afterwards laid down, every field would get its equal share of dung. *R. B.*

(*a*) We find it very difficult to get wheat cut sufficiently low, even at an advanced wage, but I endeavour to get the stubble off by harrowing, or other means, as speedily as possible after leading off the corn, and carry it into the farm yard, before it gets dried, where it becomes good manure, by mixing it with the other litter, and being pressed down by carriages and cattle before winter; were all wheat stubbles thus gathered into the yard, it would considerably increase the farmer's manure, an object well worth attending to, since manures are become so exceeding dear. *T. H.*

Shearing low is a good practice, and worthy of being adopted. *Samuel Birks Esq.*

(*b*) In Oxfordshire I have seen the stubble burnt, which seems a much better practice than plowing it in, though not either to be imitated, as the present demand for straw seems to sanction the low cutting of the crop, as practised in East Lothian. *W. Fox.*

Answer.—Laying the demand for straw out of the question, the practice of low cutting is sanctioned by the increased value of dung, independent of the additional quantity of corn gained. This the remarker seems not to have attended to. *R. B.*

I have taken every mean in my power, to prevail with my shearers to cut the corn low, but in vain; indeed where it is of great length, it is heavy enough in the hand, when cut in the common way, and if it were cut close by the ground, it would hardly be possible to wield it in handfuls to the sheaf. The labourers in Scotland may be more manageable, and the straw not quite so long as the West Riding of Yorkshire.

A Yorkshire Farmer.

Answer.—The straw in that part of Scotland, where the authors of this work reside, is as long, and the corn as heavy, as any part of the island. *G. R.*

(*l*) Without entering into philosophical disquisitions concerning the nature of lime, it is agreed upon all hands, that it renders land fruitful; the objects of the covenants, are to oblige the farmer to render his land fruitful, and to preserve it constantly in that state, till it shall revert to the landlord; it is known, that if this mode of manuring land be very often repeated, it will cease in time to have the desired effect. Experiments which would ascertain, how long the strength of lime will continue unimpaired, and how often the doze may be safely repeated, would make a valuable accession to the knowledge of agriculture; a reasonable landlord wishes to promote the prosperity of his tenants for his own sake; if he is influenced by no other motive, and would by no means bring them to an expence, which doth not appear to him to be necessary. *T. York, Esq.*

Answer.—But why should the interest of the tenant be regulated by the reason of the landlord. If the landlord wants reason what becomes of the tenants interest?

Again, if lime renders land fruitful, how can it cease to have the desired effect often repeated? The fact is, Mr York's sentiments appear to be precisely the same as those given in this report, viz. that lime will operate in certain cases, and be useless in others, therefore a covenant obliging the tenant to apply it to his ground every time it is fallowed, must by him be considered as arbitrary and absurd. *R. B.*

We are assuredly very defective in the application of lime, and the practice of laying it on upon the fallow, is continued from custom by most farmers, especially the small ones. They generally have not yard manure to cover, upon an average, one third of their summer fallows, therefore make up the deficiency with lime, because they must do something, not knowing or considering its properties or effects. Upon poor worn out soils, which have been long under the plough, lime is of little, or rather of no use, and the money expended in purchasing it, together with the labour in driving and laying it on, may be considered as in a great measure lost. *T. H.*

(*k*) I must approve of this method of laying on lime, in preference to the present practice of laying it upon fallows, except when the land is very fresh, to wit, the first fallow after paring and burning, or after the whole sward which has been lately plowed without being previously limed. *T. H.*

CHAPTER XIII.

LIVE STOCK.

THE West Riding being a great grasing district, it might be expected that much attention would be paid to selecting good breeds of stock; which, from our observation and information, was not generally the case. Indeed, the horned cattle and sheep, fed in the district, include almost all the different varieties reared over the whole island. This mixture may be attributed to the extent of the demand, which far exceeds what can be raised in the district.

SECT. 1.—*Horned Cattle.*

THE horned cattle of this district may be classed under four different heads. 1. The short horned kind, which principally prevail in the east side of the Riding, and are distinguished by the names of the Durham, Holderness, or Dutch breeds. 2. The long horned or Craven breed, which are both bred and fed in the western parts, and also brought from the neighbouring county of Lancashire. These are a hardy sort of cattle, and constitutionally disposed to undergo the vicissitudes of a wet and precarious climate. 3. There is another breed which appears to be a cross between the two already mentioned, and which we esteem the best of all. A great number of milch cows of this sort are kept in Nidderdale and the adjacent country, which are both useful and handsome. They are perhaps not altogether such good milkers, as the Holderness cows, but they are much hardier, and easier maintained. They are, at the same time, sooner made ready for the butcher, and are generally in good order and condition, even when milked. 4. Beside these, there are immense numbers of Scotch cattle brought into the country, which, after being fed for one year, and sometimes two, are sold to the butcher. Beef of this kind always sells higher in the market, than that of the native breed; and from the extent of population, there is a constant demand for all that can be fed.

Mr Parkinson at Doncaster, was of opinion, that the horned cattle of the first sort, would be much improved by crossing them with the best Craven bulls, which meets with our approbation, and is in fact practised in the interior parts of the Riding. The cattle of the Craven breed have been long famous over the whole island, and we had an opportunity, at Settle fair, to see a fine show of that sort, which afforded us particular satisfaction.

We acknowledge that the Craven cows will not give such a return of milk as the short horned, or Holderness breed, but believe this in part remedied by their milk yielding a greater quantity of butter. No doubt but that in the vicinity of large towns, where there is a great demand for milk, the latter sort is to be preferred, but in other situations, or in every place where the climate is cold or wet, the long horned breed may be advantageously kept.

A very ingenious paper upon the management of cows, in the neighbourhood of London, has been laid before the Board of Agriculture by Baron D'Alton a foreign nobleman; and, from the accurate calculations therein given, it appears, keeping cows in the house is more profitable husbandry than pasturing them in the fields, as is commonly done. During our survey of the West Riding we made repeated inquiries whether any such practice prevailed in that district; the result of which were, it was only done by a few cow-keepers in towns, who had little or no land. By a letter, received since our return, from Mr Stockdale, at Knaresborough, we were informed that this practice was common at Leeds. We therefore wrote to a gentleman there, desiring him to inquire if it was found beneficial. The following is a copy of his answer:

SIR, *Leeds, Jan.* 15, 1794.

" There are a few cows kept in the house all summer, and the way in which they are managed, is by giving them grass fresh cut, and watering the ground as the grass comes off, with the urine from the cows. The urine is preserved by a cistern placed on the outside of the cow house, and is conveyed to the land at almost all seasons, but the most profitable time for doing it is March, April, and May; by which means, and the addition of horse dung applied during the winter months, the field may be cut 4 or 5 times during the season. I am told 4 acres of land will, in this method, maintain 10 cows; and in the winter they are fed with grains from the brewers, which are very high in price, being 3s. 6d. per quarter. It will take about four pounds worth of grains to maintain a cow for the winter months, and two pounds for grass during the summer: so the expence of a cow for the whole year is about six pounds.

I kept 13 cows one winter, which were fed upon turnips and oat straw, and never got a mouthful of hay. They yielded me 30 gallons of milk per day, which, six years ago, sold upon the spot, to the retailers from

Leeds, at 5½d. per gallon. They carried it a mile, and ſold it out at 6½d. and 7d. per gallon; but it is now advanced to 8d. and 9d.

"I muſt notice to you, that the taſte of the turnip is eaſily taken off the milk and butter, by diſſolving a little nitre in ſpring water, which being kept in a bottle, and a ſmall tea-cup full put among 8 gallons of milk, when warm from the cow, entirely removes any taſte or flavour of the turnip (a).

"In the management of cows, a warm ſtable is highly neceſſary, and the currying them, like horſes, not only affords them pleaſure, but makes them give their milk more freely. They ought always to be kept clean, laid dry, and have plenty of good ſweet water to drink. I have had cows giving me 2 gallons of milk at a meal, when within 10 days of calving, and did not upon trial find any advantage by allowing them to go dry two months before calving. The average of our cows is about 6 gallons per day after quitting the calf.

"If this ſtatement affords the Board of Agriculture any information worthy their notice, I will be happy at being the inſtrument of it; and all I have ſaid is from experience. You have my ſincere wiſh for the laudable work you are engaged in being crowned with ſucceſs, and I am, &c."

In addition to the above very ſenſible letter, we may add, that one of us, for ſome years, has kept his cows in the houſe upon red clover and rye graſs during the ſummer months. They are put out to a ſmall park in the evening after milking, for the convenience of getting water, and tied up in the houſe early in the morning. One acre of clover has been found to go as far in this way, as two when paſtured. More milk is produced, and the quantity of rich dung made in this method, is supposed to compensate the additional trouble of cutting and bringing in the grass.

A variety of remarks are made upon the above letter by Mr Henry Harper farmer at Banktop, near Liverpool, chiefly from misconceiving its meaning. We apprehend few persons, except Mr Harper, thought that grains were the sole food of the cows during the winter months, which, from a second letter from our correspondent, turns out to be actually the case, as they got oat straw at the same time. We acknowledge this ought to have been noticed in the original statement, but that was not our fault, for we gave the information as we got it. After all, as no credit is taken for the dung produced by the cows, we are inclined to think the charge of keeping, will not be so wide of the mark as Mr Harper imagines.

Mr Harper seems surprised that clover grass should be cut so often, as our correspondent mentions, and thinks very little land in the kingdom is worth a second cutting. We are ignorant of the sort of land he possesses; but we have seen, in our own country, clover cut three times in one season; and, when the extraordinary manuring, mentioned in the letter, is fairly considered, the quantity of grass is by no means surprising.

What is said respecting the average quantity of milk given by a cow, after quitting calf, was not understood by us to include the whole season; as it is well known, that keep a cow as you please, she must necessarily fall off after a certain period is elapsed. Mr Harper, however, interprets it for the whole year, and gives a comparative statement of the produce, with that of a cow at Liverpool, which our correspondent's account will not warrant. The fact is, that the information communicated to us, was merely given to shew the general system of keeping cows at the manufacturing towns, and not as the result of a profit and loss account.

Mr Harper makes the following remark upon the plan we suggested for feeding cows in the house, in the summer months.

" Keeping cows in the house in the months of July and August, and in the other months when the weather is hot, is useful for either feeding or milking, and if they are well fed in the house at those times with clover, they will certainly feed faster, and give more milk; but my opinion is, that a cow, either for feeding or milking, in the summer months, if she has a reasonable supply of grass, to feed herself as she pleases, and to lay down the same in an open field, it will answer the purpose better than being confined to the house in the day, and turned out in the evening into a small bare pasture, let the water in it be ever so pure; and there is no account of the value made for the evening pasture; and an acre of clover mowed off the land in that mode, should be of three times the value of one that is grased off; or how is the farm to be carried on? for, if it is only of double value, the farm will be loosing one third part of what it should make to pay its way; and, if the dung answers to pay the extra trouble of cutting the grass, and serving the cattle, what is the difference of keeping mowing land in condition, and that of pasturing?"

As Mr Harper allows, that a cow will feed faster, and give more milk, when plentifully supplied with grass in the house, than when allowed to go at large in the fields, we are under no necessity of saying one word on this head: the question betwixt us relates to the profit of the mode now recommended, which shall be shortly discussed.

When we stated, that an acre of clover ground would go twice as far when cut, as when depastured with cattle, we were within the mark, as will be acknowledged, by any person who reflects upon the quantity trampled under foot, and made useless during wet weather. Still adhering, however, to this statement, it must be evident, if one acre goes as far as two, that the value of an acre of grass is saved, where the system we recommend is practised. Now, how is the land to be exhausted by the practice, seeing the dung of course is returned, either to the field from which the grass is taken, or to some other field of the farm, as circumstances may require. Does Mr Harper mean to say, that an acre of grass depastured with cattle, will yeild more dung, than when regularly consumed in the house? If he does not, his objections falls to the ground; and, when the additional quantity of dung, afforded by littering the cows, is taken to account, it will still be less tenable.

The practice of feeding work horses in the house during the whole season, is common in the best cultivated counties of the kingdom; and why should not this practice be extended to cattle of all descriptions? It undoubtedly augments the quantity of dung raised upon a farm. It allows it to be regularly applied to any field, according to its necessities, and prevents it from being scattered along the sides of hedges or walls, while the rest of the field is deprived of manure. In a word, if it be granted that the animal will thrive as well in the house as in the open air, (which Mr Harper concedes), a doubt cannot be entertained of its propriety.

Another gentleman remarks, "That turning out the cows, and the taking them into the house again, early in the evening, may prevent all injury of their health from confinement;" but adds, "it may be questioned whether it will not expose them to imminent danger of catching cold, and that this ought to have been enquired into." When the very persons who tried this way of keeping cows, recommended the practice to others, such an enquiry, at least on their parts, was superfluous. If the cows are tied up in an open airy house, they will not be too warm; and they ought always to be put out before the night dews begin to fall.

We have received the following information upon the same subject, from a gentleman at Sheffield:

" One of our most experienced cow keepers says, he gives 5 hundred weight of linseed dust, mixed with 3 hundred weight of bran per week, to 6 cows; others give a quarter of a peck of bran, a quarter of a peck of beans, with a peck of grains for one feed, for one cow, three times a-day. These are expensive methods, but seem to answer well, as both the cows and their owners thrive, although some persons think those feeders, who are nearest the water, thrive best of all."

Mr Bryan Waller, at Masongill says, that the expence of keeping a milch cow in his neighbourhood, (the western extremity of the Riding,) may be L. 7 per annum, and the produce L. 10. As to dairy management, many farmers bring up calves, giving them skimmed milk, after they are three or four weeks old, and the butter is mostly sent to the manufacturing towns in Yorkshire and Lancashire.

From the most minute enquiries, we did not find that the practice of keeping large dairies is customary in Yorkshire. It is principally confined to the neighbourhood of large towns, and the produce sold in its raw state, which is certainly a profitable trade.

At York and Adwalton, fortnight fairs are held in the spring for the sale of cattle; at the former for cows and oxen from the county of Durham, &c.; and, at the latter, for cows of the long horned kind, from Craven, which are chiefly in the hands of jobbers.

SECT. 2.—*Sheep.*

THERE are so many kinds of sheep, both bred and fed in this district, and they have been crossed so often, that it cannot be said to possess a distinct breed. The sheep bred upon the moors in the western part of the Riding, and which, we presume, are the native breed, are horned, light in the fore quarter, and well made for exploring a hilly country, where there is little to feed them, but heath and ling; these are generally called the Penilton breed, from the name of the market town, where they are sold. When fat they will weigh from 12 lb. to 15lb. per quarter. They are a hardy kind of sheep, and good thrivers. When brought down, at a proper age, to the pastures in the low parts of the country, they feed as cleverly, and are as rich mutton as need be. We suppose crossing ewes of this sort with a Bakewell ram, would produce an excellent breed for the low country pasture, as the Bakewell kind have exactly the properties, that the Peniston wants.

There are great quantities of Scotch sheep from Teviotdale, &c. fed in the country; numbers of ewes are also brought annually from Northumberland, which, after taking their lambs, are fed that season for the butcher. Many two years old of this kind are also fed upon turnips; and in the southern parts there are a good many of the flat ribbed, Lincolnshire sheep, which are ugly beyond description.

Upon the waste commons, scattered up and down the Riding, the kind of sheep bred, are the most miserable that can be imagined. As they generally belong to poor people, and are mostly in small lots, they never can be improved. This will apply to the whole of the sheep kept upon the commons, that are not stinted; the numbers that are put on beggar and ſtarve the whole flock. In many parts of the Riding, a ſuperior attention is now beginning to be paid to this uſeful animal, by ſelecting rams of the beſt properties, and breeds; which, it is to be hoped, will be more and more attended to.

Mr Parkinſon at Doncaſter, ſays on this ſubject, "A great part of this county is not proper to breed upon, yet ſheep ought to be kept by every farmer for improving his land; and in my opinion the moſt profitable way is to buy draft ewes in September, and to feed their lambs; after that, keep on the mothers till fat. As our ſoils are liable to rot ſheep, by floods, &c. the farmer, by this method, will not run any riſk; for if his ſheep take the rot, they will, if managed properly, be ready for the butcher at all times. The turnips upon the clay ſhould be eaten early in the ſeaſon, to make the moſt of them, and thoſe upon the lime-ſtone and ſandy ſoils afterwards, (*b*).

"I will now deſcribe the kind of ſheep proper to be bred on the ſand and lime-ſtone farms; and theſe, I think, are the Diſhley, or, as they are commonly called, the Bakewell breed; the properties of which are well known. Their wool may be conſiderably improved; it being in general of too ſhort a kind, and producing various ſorts in one fleece; viz. moſſy on the back, hairy on the thighs, or breech, and fine and ſoft on the ſhoulders and necks; which cauſes one part to be ſold at Bury St Edmonds, and the other at Halifax, to make the moſt of it.

"It is the opinion of ſome, that long wool injures the carcaſs: I do not believe it, or at leaſt it is ſcarcely perceptible; therefore I would have the wool on theſe ſheep to be of a fine combing quality, nine or ten inches long, bearing a very even top, as that prevents both loſs of wool and labour, by not having the tag end to cut.

off. The weight of the fleece to be from eight to twelve pound, if properly fed, if not, it will perhaps be only from five to feven pound. The carcafs to weigh from 20 to 25 pound per quarter with common food; extraordinary feed, from 25 to 40 pound.

" The fheep, at prefent bred in this county, I mean thefe bred upon the commons, are not worth defcribing. Their fleeces weigh from one to five pound, but very few fo much. The carcafs will feed from nine to fifteen pound per quarter—general run about twelve pound. It is my opinion, feveral thoufand pounds are annually loft in the neighbourhood of Doncafter, for want of a more improved breed of fheep.

" I think the Difhley fheep are generally too fmall: their bone and fhape are beautiful, but their fkin, or pelt, is too thin for bearing the cold (*c*). They can neither ftand the extremities of heat nor cold; and it is fometimes found neceffary to clothe them, where this breed is newly introduced (*d*). The wool of the Northumberland fheep ftands in need of great improvement: upon many of them the ftaple is much too fhort, and fome carry a hairy fort of wool, not profitable. The carcafs, though not fo inclined to feed as the Difhley fheep, yet being far larger, pays very well.

" The Durham, or Tees fheep, if improved, might pay very well; but, from what I have feen, I think little attention has been paid to them, every flock being of various forts, both in refpect of wool and carcafs. There is a number of them pretty good, but a greater number not fo. I am of opinion, a careful and knowing obferver of fheep, would raife a fine breed from the Difhley ram and Tees ewes. Sheep are an animal difficult to bring to perfection, as both wool and carcafs are to be attended to; but one thing I am clear in, that the best carcasses will produce the best wools; like as good land affords good grain."

An intelligent farmer, for whose sentiments we entertain great respect, gives us the following account:

"The sheep that are kept in this extensive county are as variable as the soil and climate, and in some degree suited to each. Most of them have made, and are yet capable of great improvement. Those bred above Peneston are well adapted to those uncultivated barren mountains, where they have little to feed upon but ling or heath, and are perhaps the least capable of improvement of any other: but as you have seen them, I need not be particular in describing them. I imagine their fleece, taking ewes, wethers, and hogs together, will average about 2½ or 3 lb. which will be worth 2s. 6d.; of late years a little more. Those bred upon York wolds are very numerous, and far the best in the county. It being a dry, flinty, lime-stone soil, and capable of cultivation; by growing turnips for their winter support, they raise some of them to good weights, 27 or 28 lbs. per quarter when fatted. Those farmers occupying large districts of land, can keep great flocks, which makes it worth their attention to improve them, and great improvement some of them have made by crossing with Bakewell's rams, and breeding from the best Northumberland ewes. This has rather decreased the weight of the fleece, but improved the staple, and given them a property to feed much quicker and fatter. Those sheep will weigh when fat, from 14 to 28 lb. per quarter, in proportion as they are supported with food; and the fleece upon the best walks will average 6 lb. or better, which this year is worth about 4s.; those on the poorer walks from 4 to 5 lb. worth from 2s. 6d. to 3s. 2d.

"What are bred in this neighbourhood upon waste

grounds are of ſmall conſequence. They are the worſt in the county, being bred from all ſorts; and belonging chiefly to poor people, in ſmall lots of 10, 15, or 20 each, will never be bettered till the lands are incloſed. We have a few gentlemen farmers begun to breed from Northumberland ewes and Bakewell's rams, which I think, makes far the beſt and moſt profitable ſtock; but for want of room, nothing of conſequence can be done here in the breeding line. The chief practice of our farmers is to buy ewes at Peneſton, or from York wolds, or Northumberland, at Michaelmas, fatten the lamb in the ſpring, and the ewe afterwards, changing every year. Being near a manufacturing country, full of opulent tradeſmen and merchants, lamb always bears a good price, being worth 6d. per lb. nearly, on an average, all ſpring and ſummer. Where there is room to breed a few of our own beſt ewe lambs every year of the above ſort, to keep up a ſtock in proportion to the ſize of the farm, I believe it moſt profitable, as ſtock bred upon our own ſoils, if of a proper ſort, will fatten their lambs and themſelves too, much ſooner than thoſe brought from any other part. The fleece of theſe, where gentlemen have brought them to tolerable perfection will be 6 lb. average, and worth 4s. 6d. or 5s."

Another farmer of great profeſſional merit, and intimately acquainted with the ſheep huſbandry of the Weſt Riding, has obligingly favoured us with the following communication:

"The Diſhley breed of ſheep, are moſt certainly gaining ground every where in the ſouthern, and eaſtern parts of the Riding. Rams of that kind, are far more ſought after than heretofore, and I am in no doubt of their becoming the eſtabliſhed breed of this county.

"Our mode of managing them is this: The ewes have turnips previous to their lambing, which generally is about March, when we take ſuch as we mean for rams with their dams, to better paſtures; the reſt to ordinary keeping. About one fourth of them produce double burthens, generally ſmall, but exceedingly inclinable to be fat, even from their firſt appearance, (if in any tolerable keep,) as the ewes are bad nurſes. We clip the latter end of May, or beginning of June; take the lambs from them fore end of July; milk the ewes twice or thrice to eaſe their udders; put them into the bareſt paſtures we have till Michaelmas, after drawing out ſuch of them as are moſt diſapproved of. Theſe being put to the beſt paſtures afterwards, (if theſe failed,) to turnips or rape; ſold at Chriſtmas, generally at Wakefield, for from 40 to 50 ſhillings each, and fat enough: the lambs are put to the beſt meat we can ſpare, but moſt generally to old paſtures, and eddiſh if we can, till about November, when they go to turnips; the wethers to the beſt paſtures after turnips, which make very fat by Auguſt following; have ſold ſeveral years back my ſhearlings, at 40 ſhillings per head, laſt year 50 ſhillings, at Wakefield; thick fat, no lumberly weights, from 20 to 22 lb. a quarter, neat ſmall fine bone, fine grain and fine colour; and worth more by a penny per lb. than any large boned mutton in the kingdom, though not generally ſold for it at preſent. Should be glad to hear of any other ſort of ſheep which get ſo fat, and worth ſo much money at 17 months old, notwithſtanding the difference of the quantity of food eat by thoſe, and the other long wooled breeds, which I am convinced is very great. The rams eat nothing in winter but turnips or hay, and graſs or clover in ſummer; no need of *oil cake* or *corn* to make theſe *thick*, *fat*, and *handſome* for ſhow, as is, I am informed, indiſpenſably neceſſary to all the other long wooled breeds. They are fit for any wholeſome ſoil, on a temperate climate, and will moſt certainly pay more upon thin poor land than any other kind: nor am I in the least doubt of their being much superior upon the very best.

" A particular friend and neighbour of mine, the year before last, wintered 100 of these ewes in the straw fold, which kept the produce of two threshers down, better than 20 beasts would have done. He gave them a third part of a common cart load of turnips every day, to keep their bodies open, as the straw would otherwise bind them. This winter he has them come up every night of themselves, which shews they like it; they eat the straw very greedily, and goes out of themselves in the morning to an adjoining grass field. This change of food and warm lodging, agrees with them very well to all appearance, the grass having the same good effect as the turnips, and the straw in the night time, more agreeable to their nature than confinement. But the manure being subject to heat the sheep, when too great a quantity is accumulated together, it should be led out of the fold when that happens to be the case. He led out about 150 loads of manure at Christmas, chiefly from this fold, which he says is in as fine a state of fermentation, as any he ever had. He disapproves of swine, or any other stock being with them. This is a hint, I hope you will not think unworthy of remark, if only a substitute for cattle, when they are now scarcely to be had at any price. This person, notwithstanding his manner of wintering, gives them turnips previous to their lambing, to increase their milk; but summers them upon a high poor gravelly soil, upon seeds of one, two, and three years lay; yet both ewes and lambs are fat. I have heard of several tenants, who, before his time, upon the same farm, could not live upon it, but lost much money; and, from this gentleman's peculiar and most commendable mode of management, I am certain the consequence must be diametrically opposite."

From an anonymous paper, transmitted to us, we select some further information on this important subject.

"The advantages of sheep are numerous, but the most beneficial sort is the Dishley breed; a man of knowledge, may put any kind of wool upon them he chooses, according to the soil; and their carcases may also be improved in a similar manner. They will also pay better for the food they eat, than any other of the numerous breeds that prevail in the Riding.

"Sheep improve land more than any other animal, and I account for it in this manner. They have a small mouth, and eat leveller by consuming all kinds of weeds except thistles and nettles. They tread the ground in a gradual but continual manner, by which they fasten the earth, and do not break the swarth, or bruise the plant in wet soils. By gradually treading the land, the superfluous water is pressed out during wet weather; and, in dry weather, the drought is thereby prevented from getting in."

A Yorkshire farmer says, "the sheep kept on commons, might be much improved, if several of these small breeders would join, and hire a ram of a right sort. There is an act of parliament, called the cultivation act, which prohibits rams from running on wastes, from the 25th August to the 25th November, every year. If this act was strictly put in force, these little sheep breeders would soon unite, and hire a ram for their mutual interest, as any breeder would furnish them one at a low rate, rather than have the neighbourhood over-run with rams of a spurious race. I believe the act above mentioned, imposes no penalty; but if there were penalties imposed, it would be an excellent method of preventing this enormous evil."

Sect. 3.—*Horses.*

There are not many horses bred, except in the eastern parts of the Riding. The size of those employed in the western parts, is generally small; but they are hardy, and capable of great fatigue. In other parts of the Riding, they are large, and sufficiently able for any field operations. Those used in the waggons are strong and well made.

A farmer in the West Riding, on this head says, "In respect to horses, very few are bred in this neighbourhood, scarcely any for sale. The farmers and manufacturers breed a few for their own use; as such every man gets of a sort that is most likely to be adapted to his own business; some galloways, worth, at 5 years old, from L. 10 to L. 15; some half bred horses, fit for either plough or saddle, about 15 hands, worth, at 5 years old, from L. 18 to L. 25; and a few of the heavy black ones, which will be worth from L. 25 to L. 30, if free from blemishes: those will get to 16 hands high. But the East Riding is the circuit for horses: there the best road and coach horses are bred in England, and of any price almost, from 20 to 60 guineas at 5 years old. This circuit is by no means adapted to the breed of horses.

Horses and Oxen for draught.—Very few oxen are wrought in the West Riding; and these only upon the farms of proprietors. We know working of oxen is a popular topic; but, from what we could learn upon this subject, the practice is not likely to become general. From their being almost universally given up, in those places where they were formerly in repute, a suspicion arises that working them is not attended with profit. Those who object to the use of oxen say, that there is nothing faved by working them, as the difference betwixt the value of a horfe and an ox, when unfit for work, is more than compenfated by the fuperior labour of the former when employed. At the fame time, it is a bufinefs of infinite difficulty, to get perfons to work them (c).

The working of oxen, in preference to horfes, is a queftion which has often been difcuffed, and many plaufible arguments have been adduced in favour of the former. "What," fays the theorift, "can there be any comparifon betwixt the two animals in refpect of profit? You buy the ox cheaper than the horfe, you fupport him at lefs expence, and finally, when he is ufelefs for work, you make him up for the market, and fell his carcafe for more money than he was worth when working in your team; whereas, the horfe is a coftly animal at the outfet, muft be pampered with plenty of corn and hay, is expofed to many diforders, and at the laft is only a dinner for a dog." All thefe things may be true, and yet the horfe may prove the cheapeft of the two for carrying on farm labour.

We have already noticed, that the giving up of oxen, and fubftituting horfes in their ftead, affords an argument, *a priori*, in favour of the latter. In Britain, oxen were in former times almoft univerfally employed in tilling the ground, and they were gradually laid afide as improvements were introduced. This is a fact which will hardly be queftioned; for, at this day, except in remote uncultivated parts, there is hardly an ox team employed, unlefs it be on the farms of landed proprietors, who probably have been induced to ufe them from public fpirited motives, without enquiring into the practical refult of their operations. They have been told, that it is owing to the obftinacy and ignorance of the farmers, that oxen are not generally employed in farm labour; and, that to remove thefe obftacles, no method would be so salutary as to work them upon their own farms. That this is a fair account of their motives, we presume, will hardly be disputed; and that the fashion, promoted and recommended by the proprietor, was not adopted by the farmer, must solely be attributed to his conviction, that the working of horses was not for his interest.

That this is actually the case, we shall endeavour to prove. The very strongest ox will not do the same work as a horse. He cannot be drove at the same step, nor will he work in like manner from day to day. He not only does less labour when employed, but must be refreshed with more rest, or else he would soon turn unfit for work altogether. This of course requires two pair of oxen, to do the work which will be performed by one pair of horses, and nearly the same capital stock will be necessary in both cases. Instead of maintaining two horses, you have four oxen to support, which certainly turns the scale. "Oh," but says the theorist, "a little straw will serve for the ox, whereas your horse cannot live without hay."—Straw for a working ox! very good indeed. If you work him like a horse, he must be maintained like a horse. He must have his hay and his turnips, and possibly his corn also, if he is kept at hard work. That working oxen are not always supported in this manner, we chearfully grant; but how are they wrought? In many places six, eight, even a dozen, are yoked in a team: We here speak of the northern parts of Scotland, where oxen are more generally used than in any part of the island. In a word, oxen cannot be used for dispatch like horses, and, in critical seasons, when there is a necessity for a push, the value of the stock might be lost upon a single crop. The only point in favour of oxen, is their value at the latter end. Here on comparison can be made. Notwithstanding which, we adhere to our first opinion, that this is more than compenſated, by the difference betwixt the value of their labour when employed. Indeed the ſentiments of the greateſt part of practical agriculturiſts coincide with thoſe we here give, upon this branch of rural œconomy.

SECT. 4.—*Hogs.*

HOGS of various breeds are kept, and they have of late received much improvement. We never could diſcern the profit of them to the farmer in any other view, than when they are fed upon the offal of his corn, which is for no other uſe. If their number is proportioned to the ſize of the farm, a conſiderable advantage may be derived from keeping theſe animals, and they may be carried on during the ſummer months, by giving them cut clover and vetches, which will ſwell their ſize, and prepare them for fattening upon the refuſe corn.

SECT. 5.—*Rabbits.*

THERE are not many rabbit warrens in the diſtrict, nor indeed much ſoil of a proper kind for that animal. It is only upon ſoſt waſte lands they ought to be ſuffered to remain, as, upon cultivated land, they are a perfect nuiſance.

SECT. 6.—*Poultry.*

THE profits ariſing from this article, are of no importance in an agricultural point of view; for it may be queſtioned, whether the expence of ſupporting them, when added to the damage they do to houses, and the depredations they commit on corn, both at seed time and harvest, does not far exceed any benefit which may be drawn from keeping them. We allow it is very convenient for a farmer, to keep a few for his own table, and to supply his family with eggs; but any greater quantity we maintain to be prejudicial to his interest.

It is really diverting to read the modern declamations against inclosures, and the increased size of farms. The authors alluded to, take it for granted, that these measures lessen the number of poultry, and that the only way of getting the markets plentifully supplied with that article, is to lessen the size of farms, and to keep the waste lands of the kingdom in their present unproductive state. At this time we shall not enter upon these topics, being convinced that such a discussion is wholly unnecessary. We may only say, that where poor people, labourers or others, get poultry supported at the expence of the farmer, it may be a material object to them, seeing that they are fed by others; but, considering the question, so far as respects public advantage, the breeding and feeding of poultry ought never to be ranked as an object deserving the farmers attention.

It might also be a question, whether the benefit said to be derived by poor people is not in many cases imaginary. We have heard, that in some places, (not in the West Riding), a man would spend a day in going to market to sell a pair of chickens, the value of which did not compensate for the loss of time spent in disposing of them.

SECT. 7.—*Pigeons.*

If poultry is not beneficial to the farmer, pigeons are far less so; nay, they are a certain loss to every farmer, who has land contiguous to where they are kept. Pigeon houſes in general belong to landed proprietors, and if they are poſſeſſed by farmers, a rent of courſe is put upon them. It is impoſſible to calculate the loſs ſuſtained by the public at large from this voracious and deſtructive animal; and we conſider it would be of great utility, to diſcountenance their increaſe, by impoſing a tax on every houſe where they are kept, in proportion to its ſize.

Whether the farmer has a right to ſhoot pigeons, when committing depredations on his property, is a queſtion which has been diſputed in ſeveral parts of Britain. To us it appears clear, that if he has not ſuch a right under the preſent laws, he ought inſtantly to be inveſted with it. Shall a man be baniſhed when he ſteals a certain part of my property, and hanged when he takes a larger portion, and muſt I patiently ſubmit to greater depredations, merely becauſe they are committed by a pigeon? What is it to me, whether the owner of the pigeon takes my property with his own hand, or keeps theſe animals to pigeon me out of it? The law protects me in the one caſe, and certainly ought, and probably does, protect me in the other alſo.

Several attempts have been made in the northern parts of the iſland, to puniſh perſons who ſhot pigeons, which in general proved unſucceſsful. Some old obſolete laws have, in theſe caſes, been founded upon, which are a diſgrace to our ſtatute books. The matter has not as yet, to our knowledge, received a fair inveſtigation, ſuch complaints being uſually ſet aſide upon previous points, or diſmiſſed, becauſe the complainer had either no legal right to keep pigeons, or could not indentiſy his property. As for our parts, we decidedly think, that no man can have a juſt right to feed his live ſtock of any kind, upon the grounds of another; and, that where pigeons are kept, the owner ſhould either confine them in the houſe during ſeed-time and harveſt, or ſubmit to their execution upon the ſpot, when they are allowed to fly about at large, ,and deſtroy the corn of other people, at theſe important ſeaſons.

Sect. 8.—*Bees.*

We don't think many bees are kept in the Weſt Riding; at leaſt the information communicated to us incline us to believe they are a ſcarce article. Perhaps the ſevere winters, and cold backward ſpriugs, which we experience in this iſland, are inimical to this induſtrious little animal. After all, the ſubject cannot be conſidered as very intereſting to the farmer, however beneficial to particular individuals.

NOTES on *Chap.* 12.

(*a*) This is a valuable fact, now pretty generally known and practised. *M. Culley.*

(*b*) There are certainly many hundred acres betwixt Bautry and York, in open fields, capable of raising as good sheep as can be bred; there is no doubt, also, much land not proper for that purpose; but if all the fields were inclosed, this part of Yorkshire would cut a more respectable figure, than it does at present, being torn in pieces, or rather turned over by half starved farmers, and half starved horses, till the crop is hardly worth the reaping. *A Yorkshire Farmer.*

(*c*) I conceive this assertion to be unfounded; as experience proves that the Dishley sheep will bear either heat or cold, as well as any other breed in this Riding. *A Farmer.*

(*d*) I am sorry to contradict my friend Mr Parkinson, but I never heard of any of these most valuable sheep being cloathed, except those of the highest estimation, and in the hands of the oldest breeders, and rams which are let for the season for from 100 to a 1000 guineas each. Surely these are well worth 2 or 3 yards of flannel; I know sheep bred with attention for many years, from this invaluable sort, which want no cloathing, and which certainly pay more for what they eat, than any sheep the world ever produced. An experimental farm, under the direction of the Board, I approve of much. *A Yorkshire Farmer.*

If the best sheep produce the best wool, surely the Dishley sheep must; but as the carcase is nine times the value of the fleece, surely it claims the first attention. *A Yorkshire Farmer.*

(*e*) This difficulty may be a valid objection to the individual who must consult his own interest, but does not apply to the abstract question, whatever way that should be determined. The farmer at present is in general induced to prefer horses, more by a spirit of gambling and speculation, than a regular calculation of loss and profit One man sells a lucky colt at a high price, and all his neighbours buy mares to work with, in hopes of obtaining ſimilar high prices; may not the powers of oxen be depreciated, not only from our ignorance ariſing from diſuſe, but alſo to the uſual mode of employing them. They are taken from work, and fed till fat, and their place ſupplied by the riſing ſteers. If any raw colts were worked, horſes would ſoon loſe their reputation; a working ox ſhould be kept till his powers begin to fail, or to the age, after which it is found he cannot be fattened.

The above note is taken from a copy of the Survey, wherein the names of Meſſrs Sheldon, Pulkine, and Mitchel are marked on the title page.

Anſwer.—However, juſt the arguments may be, that are uſed in favour of working oxen, ſurely the reaſons given why farmers prefer horſes are frivolous, chimerical, and abſurd. *R. B.*

CHAPTER XIV.

RURAL OECONOMY.

SECT. I.—*Servants, Labourers, &c.*

THE West Riding being a great manufacturing district, it may at once be inferred, that labour of all kinds is higher than in those districts where manufactures are not extensively carried on. From the result of our enquiries it appeared, that wages varied considerably, even in the district itself; but, that in most cases they were highest in the neighbourhood of the manufacturing towns, and that for these some years past, they have greatly increased.

We suppose the wages of a house servant (of which kind as already said, most of the ploughmen are) may be estimated from L. 25 to L. 30 yearly, including maintenance. There is a practice which prevails over a considerable part of this district, of giving them drink both forenoon and afternoon, be the work what it will; which is a ridiculous custom, and ought to be abolished without loss of time. What can be more absurd, than to see a ploughman stopping his horses half an hour, in a cold winter day, to drink ale (*a*)? We suspect the practice is so deep rooted, that it will not be easily removed without a compensation (*b*). This ought to be done at once, as being an encouragement to idleness; and, from wasting much time, a great obstruction to improvements.

The hours of labour are generally in summer from six to six, with the usual time for rest and refreshment, which gives betwixt nine and ten hours labour each day, and in winter from light to light. Much of labour, such as ditching, hedging, threshing, &c. is done by the piece, but the prices vary greatly in different places. We only add, that when the farmer is a proper judge of his business, piece work is not only most to his advantage, but the only way by which an active diligent servant can be properly rewarded for his labour.

Upon the article of wages, the following paper is sent us by William Payne, Esq; of Frickly, near Doncaster.

" One word for the labouring peasantry.—Throughout this work, and almost every other of the kind, there seems a kind of complaint of the high rate of wages, in rural labour. Now, as the landlords can speak for themselves, as the clergy can speak for themselves, and as the farmers can either do it, or get others to do it for them, it is but reasonable that the poor labouring peasants should have something said for them. I believe the *fact* is, that the labouring peasantry never had greater difficulties to incounter in the rearing of families, than they have at present, notwithstanding the *apparent* high rate of wages; for, that it is apparent only, will be evident to every attentive observer of the case. During the course of the present century, the landlord has trebled his rent, the clergyman or lay rector, has doubled his tythe, the farmer has increased his property, and maintained his family in conveniences and comforts, at least *decent*; but have not the poor's rates increased enormously, incontrovertibly shewing the low condition of the poor. I do not pretend here to examine the many ingenious reasons that have at different times been adduced to account for it; but this is the fact: It would be curious to develope the simple causes of the present situation of things between the *farmer*, who in an enlarged view, must be considered as the agent or steward of the other orders, and the laborious peasant, who must do *all* the work. In the first place, what has enabled the farmer to pay the landlord and tithing-man, lay or clerical, the mighty advance of rent and tithes? As all improvements in cultivation are produced by an immense increase in *labour*, they alone do not satisfactorily answer the question; No: the true reasons for this ability of the farmers are, the high rate of his *products*, and the comparatively *low rate* of *labour*. I know many superficial observers will exclaim, at what will appear to them the absurdity of this solution of the question; but when we shall have gone a little farther into it, we shall perhaps all be convinced there is not so much absurdity as may at first sight appear. It will be said, have not wages been at least *doubled* in the time you mention? Though they may have been doubled, has not the price of necessaries of nearly all kinds been doubled, some nearly trebled, and some of the more immediate necessaries for a young family, as milk, &c. in winter, can scarcely be procured for money. In addition to this, the present mode of taxation on *consumption* bears almost *exclusively* on a poor man with a *large* family, for his *whole* income must be spent in necessary consumption; and our Premier says, the revenue takes four pence from every shilling of the labourer's pittance. This circumstance opens to me a clear view, (and I wish in my conscience I could place it in such a light as to convince every man of property in the nation) of the real causes of the continued poverty of the labourers, notwithstanding the increase of wages; for taxation of articles of consumption, must necessarily, though circuitously, raise the price of the article, and thus fall with double and destructive pressure on the man who is placed in the situation of father of a large family of children, with nothing for their maintainance but the earnings of his daily sweet and toil. This mode or system does not take from a man in proportion to his *ability*, but in proportion to his *inability*—a melancholy conclusion. No wonder that bastards should encrease. The young man has a just dread of marriage under these circumstances, as by it he well knows he changes a life of *ease*, *plenty*, and *independence*, for one of *distress*, *want*, and *slavery*, if a young family should be the consequence.

"About 60 years since, my grandmother gave from 6d. to 9d. per day to her threshers in winter.—She bought good beef from 1s. 5d. to 1s. 6d. per stone, of 14 lb.—oats from 6s. to 10s. per quarter.—Old milk at ½d. per gallon, new ditto 1d.—Butter from 2d. to 4d. per lb.—Malt from L. 1 to L. 1 : 5s. per quarter, and other necessaries in proportion. At the present time (1794) from 1s. to 1s. 3d. is given to a thresher in winter.—We buy good beef from 3s. 6d. to 5s. per stone, of 14 lb.—Oats from 22s. to 30s. per quarter.—Old milk not to be had in any quantity at 2d. per gallon, new milk at 6d. to 8d.—Butter from 7d. to 14d. per lb.—Malt L. 3 per quarter, and most other necessaries at a triple rate compared with the above period. No one, after a candid comparison of these periods, in regard to wages and provisions, can in his conscience, (if he has any), think that the high rate of wages is the real cause of complaint."

The same gentleman in a subsequent letter says, "Since the time I wrote you last, *existing* circumstances have so ordered it, that the poor in this Riding, partly from the increase of wages, and partly from the decrease in the price of corn, &c. must be allowed to be in a much more comfortable state than they were in at that time; yet, on the whole, I remain under the conviction, that *our* fyftem of their management is impolitic and inhuman, and that *your* method of paying them in corn, &c. is quite the reverfe."

Although we approve of the general principles laid down in the above paper, and applaude the anxious defire which Mr Payne difplays, to meliorate the fituation of the labouring peafantry, yet we cannot go fo far as to admit, that his arguments are wholly incontrovertible. If an average is taken of the prices of grain, during this and the laft century, and a fair ftatement] made of the rate of labour during thefe periods, it will be found, that the latter has rofe much more in proportion, than what produce has done. We are rather inclined to attribute the diftreffed ftate of the labouring peafantry, to their mode of living being in a great meafure changed from what it was in former times; and Mr Payne would have been in the right, if he had faid that wages had not kept pace with the change that has taken place in manners. Again, we muft impute the increafe of the poor's rates to the fame caufe, and not to the low rate of wages; which is demonftrable from the greateft rife of the rates taking place in the neighbourhood of manufacturing towns.

We have heard of many propofals for regulating the rate of wages, but are totally adverfe to fuch a meafure. Thefe propofals are never meant to ferve the lower ranks, but folely to keep them down, which in a free country is arbitrary and unjuft. If the rent of land was previoufly regulated, the price of provifions, and confequently the rate of labour, might admit of fuch regulations; but, before the firft is accomplifhed, the others cannot with juftice be attempted. We believe it is beft to leave things of this nature to their ordinary courfe, and like water they will in every cafe find their proper level.

The only way that we know of for making the labourer's wages proportional to the rise or fall on the value of money and provisions, is to pay him in *kind*; that is, with a certain quantity of corn, as parties shall agree, which insures him, at all hazards, a comfortable subsistance, and prevents him from a daily or weekly visitation of the markets. When the labourer is paid in money, it exposes the thoughtless and inattentive to many temptations; whereas, when paid in kind, he cannot raise money to gratify the whim of the moment. In those counties where this mode of payment has been long established, we believe ploughmen and labourers are on the whole better fed, live more comfortably, and rear healthier children, than in those parts, where, from being paid in money, the currency of the article facilitates the expenditure, and prevents him from laying by a stock of provisions for his support, when laid off work by casualties or distress.

In the county where we reside, nearly the whole of farm servants are paid in the manner we are recommending. They have a certain quantity of grain; maintenance for a cow summer and winter; a piece of ground for planting potatoes, annd raising flax; and whatever fuel they require, driven gratis. These, with the privilege of keeping a hog and a few hens, enables them to live, and bring up their families in a comfortable manner; and, while their income is considerably less than people of their station in England, they are on the whole better fed, better dressed, and enabled to give a better education to their children. Placed under these circumstances, they are a respectable set of men; and for frugality, faithfulness, and industry, they will bear a comparison with their brethren in any quarter. We therefore anxiously recommend the introduction of a similar mode of paying farm servants into the West Riding; which, although it might at the first be attended with ſome difficulties, would contribute to the public good, and to the advantatge of the labouring peaſantry in many reſpects.

SECT. 2.—*Price of Proviſions, and Landed Produce.*

As the Weſt Riding, from the extent of population, is unable to ſupply itſelf with proviſions, the prices are full as high as in any part of the iſland. From the information procured it appears, that though in general no ſcarcity is experienced, yet, in particular ſeaſons the price of grain has riſen to an extraordinary height. At Wakefield market in July 1795, wheat was ſold at the enormous price of L. 9 per quarter; and it may be remarked, that during ſuch critical periods, the country which depends upon foreign ſupplies, muſt comparatively pay much higher prices for the articles which cannot be furniſhed within its own bounds, than what they do in ordinary ſeaſons; and that prices muſt neceſſarily advance to a far higher rate than is uſual in thoſe counties where the articles are produced. The ſcarcity is there felt in a ſerious way, and it requires great exertions to provide a ſupply, which was evident from the unlimited powers given at the time above mentioned to thoſe perſons appointed, from the manufacturing towns, to purchaſe grain.

It is unneceſſary to give a ſtatement of prices of proviſions during the time we remained in the diſtrict, as, from the fluctuation of markets, no light would thereby be thrown upon the value of produce. We may only hint, that the cheapeſt article of proviſions was poultry, the cauſe of which we attribute to the taſte of the inhabitants, who very judiciouſly give a preference to well fed beef and mutton, which is furnished them in the greatest perfection.

We have noticed the high price of wheat in summer 1795, which was doubtless a serious and alarming evil, and proceeded from a real scarcity of that grain over the greatest part of Britain. But, does the farmer in general receive greater prices for his commodities than the rates of rent and labour entitle him to? We answer in the negative; for both have advanced in a much greater degree, than any rise which has taken place in the value of produce. This must be attributed to the impolitic regulations of the legislature, which in fact combine to depress the agriculture of the country, by obliging the grower of corn to sell it at certain rates, whether he is able to do so or not. When there is a demand for what, in the general acceptation of the word, is called manufactures, and prices rise, it is immediately taken for granted, that the country is in a flourishing state; but the moment corn, (which strictly speaking is the first of all manufactures), sells briskly, and prices get up, the hue and cry is raised, and every exertion is used to bring in supplies from those parts, where, from lowness of rent, labour, and taxes, it can be afforded at one half of the price.

It must not be thought, that we are here contending for high prices of grain as necessary to a flourishing agriculture, or that we would wish to depress the manufacturing interest of the country. No, we only desire that each should have fair play, and that the one may not receive a preference to the other. If protecting laws are necessary for the welfare of the farmer, as all our corn laws since the Revolution have supposed, let them be rigorously adhered to; and as they were made for his encouragement, and upon the faith of them he probably made a bargain for his farm, let them not be suspended because required by the capricious diſpoſition of manufacturers. In unfavourable ſeaſons, how is he to be compenſated for the deficiency of his crop, but by receiving greater prices than uſual for what he carries to market? and when he enjoys this right, he enjoys no more than what is actually poſſeſſed by the meaneſt manufacturer in the kingdom. The manufacturer indeed is ſtill further favoured. What with prohibitions, and duties on foreign goods, he may be ſaid to enjoy the home market without a rival; and the farmer muſt of neceſſity purchaſe ſuch of his commodities as he ſtands in need of, even allowing he can buy them at a lower rate from a foreign merchant. It is therefore but fair and equal, that the laws ſhould give the farmer a ſimilar encouragement in the ſale of commodities to the manufacturer, unleſs during the times of real ſcarcity, when the public ſafety requires private intereſt to be ſacrificed. The ſubject ſhall be further elucidated under the head of Corn Laws.

SECT. 3.—*Fuel.*

THIS moſt neceſſary article is in general plentiful over the whole Riding, and, in a comparative view with other diſtricts, is ſold very cheap. In thoſe parts where any ſcarcity prevails, they can be ſupplied without material inconvenience, by means of the numerous rivers and canals which interſect the whole diſtrict. It was ſuggeſted to us by a gentleman at Settle, that where a ſcarcity prevailed, it might be remedied by Lords of the Manor making trials to diſcover coals, and by holding out rewards, or granting favourable leaſes, to perſons willing to adventure in ſuch undertakings.

NOTES on Chap. 14.

(*a*) This is certainly a moſt abominable practice, but from long eſtabliſhed cuſtom, I cannot deviſe how it can be remedied.

T. H.

(*b*) This is a bad cuſtom, but how it is to be aboliſhed I cannot tell. *A Yorkſhire Farmer.*

Anſwer—The remedy is already ſuggeſted in the text: Let the value of the ale be paid to the ſervant in money, which probably would be as much for his intereſt, and certainly more advantageous to the farmer. In thoſe places where long yokings are taken, ſay ſeven or eight hours, it may be neceſſary to feed both men and horſes on the ground; but this practice we cannot recommend, unleſs in urgent caſes, it being very injurious to their health. In the beſt regulated agricultural counties, five hours labour in the morning, and four hours in the afternoon, when the ſeaſon allows, and five hours, or five hours and a half, in ſhort days, is conſidered to be as much as horſes are capable of ſuſtaining, and yokings of this duration require no refreſhment on the ground.

R. B.

CHAPTER XV.

POLITICAL OECONOMY AS CONNECTED WITH, OR AFFECTING AGRICULTURE.

SECT. 1.—*Roads.*

THE utility of good roads is at firſt ſight ſo evident, that we need hardly ſay this ſubject deſerves particular attention. In the Weſt Riding, there are a great number of very good roads, and likewiſe a number that are indifferent. From what we could learn, they are generally under good management, and the funds well applied. In many places of the diſtrict, particularly near the manufacturing towns, materials are bad. To this circumſtance, more than any impropriety of management, we attribute their inſufficiency. At the ſame time, the ingenuity of the ſurveyors was conſpicuous, in burning free ſtones and brick, to ſupply the want of harder materials.

As theſe burnt materials make at the beſt but a very imperfect covering, and need to be frequently repeated, it appears to us, that hard ſtones might be brought, by water carriage, from the more eaſtern parts of the diſtrict. This might probably be expenſive at firſt, but we are convinced, would be found cheaper at the long-run. From Haliſax to Wakefield, the road is in the moſt miſerable condition; and if it was so when we travelled it, in the end of October, it must be nearly impassable during the winter months. This is a very public road, and no expence ought to be spared, to render it good and sufficient.

We apprehend, the weight of the numerous waggons that pass over this, and other roads in the manufacturing part of the county, must always render them bad, so long as they are repaired with soft materials. We saw some roads, that had been newly covered with burnt stones and bricks, crushed down at once by the weight of these carriages: let us suppose rain to fall, and remain in the track or rut so made; another waggon comes, and cuts down still further; and a third puts them in as bad condition as before they were repaired. By these waggons, an endless expence is created to the public, and still bad roads are the consequence.

There was nothing gave us greater satisfaction, than the paved foot-paths upon the sides of most of the roads in the manufacturing part of the country. This shews an attention to the comfort of foot passengers that is very laudable. We have noticed in the Journal, these foot paths are also made "bridle roads;" a practice which can only be excused by the peculiar badness of the main road.

The roads are a very heavy article of expence to the farmer; and here, as well as in most other parts of the island, the burden is chiefly laid upon the occupiers of land. It cannot be properly called a part of the rent; as, if the work is rightly laid out, full value is received from it: the farmer travels the road with more ease and convenience to himself; and is enabled, from the improvement made by his labour, and money, to carry more corn to market, and to return with a heavier load of dung, than he could do if the roads were in their natural state. Road expence, therefore, cannot be viewed in the ſame light as tithes and poors rates; theſe two articles being conſidered by every farmer as a part of his rent, and not as given for value received.

It has often appeared ſurpriſing to us, how the ſupport of the bye-roads ſhould be thrown upon the poſſeſſors of land; and perſons of almoſt every other rank allowed the benefit of them, free of all charge whatever, or at the moſt paying only as houſeholders. In many caſes, thoſe who pay leaſt for the making good roads, have the greateſt ſhare of the profit (*a*). The turnpike laws are not founded upon ſuch falſe principles, but every perſon by them, is obliged to contribute his ſhare of the expence for ſupporting the roads, in a direct proportion to the uſe and benefit he receives from them.

The ſtatute labour paid by the farmer for the ſupport of the roads, is ſix days labour of a team with three horſes, or four oxen and one horſe, and two able ſervants, for every L. 50 of rent, or leſs or more proportionally, together with an aſſeſſment in money of 6d. per pound upon the rent, or higher if the juſtices ſee neceſſary. Statute labour is alſo paid by the inhabitants and occupiers of tenements, woods, tithes, and hereditaments. The ſurveyors are nominated annually, upon the 22d September, at a meeting of the inhabitants of each pariſh or townſhip, who make up a liſt, not exceeding ten perſons, whom they think fit for that office; which is given in to the juſtices, who appoint one or more out of the liſt, as they ſee neceſſary. The ſurveyor or ſurveyors collect the aſſeſſment, ſee the work properly executed, and, when their time in office is expired, they lay their accounts before another meeting of inhabitants, and afterwards before a juſtice of the peace, who may paſs, or poſtpone them to the ſpecial ſeſſions, to whom every perſon who thinks himſelf aggrieved may appeal.

In making up the liſt of ſurveyors, the inhabitants

place the perſon they wiſh appointed, firſt, and the juſtices generally appoint him accordingly. If the ſurveyor is deficient in his duty, he is fined in a ſum not exceeding L. 5, nor leſs than L. 2 for every neglect; and as he muſt produce his accounts at a veſtry meeting, he can hardly eſcape if culpable. The auditing the accounts annually is a very proper ſtep, and prevents that diſorder and confuſion, which has been well known to have taken place in ſome other counties.

As great complaints prevail over the whole kingdom againſt the adminiſtration of the bye-roads, we are clearly of opinion, that ſtatute work in kind ought to be aboliſhed, and the value thereof paid in money, which would be a meaſure of great public utility. It is an old ſaying, though not the leſs true on that account, " that one man may take a horſe to the water, but a hundred men will not make him drink;" and the ſame thing will be found applicable to road work, when performed by the perſon who is liable, unleſs he accounts himſelf intereſted in the application, which is nine times out of ten not the caſe. Beſides, it is abſurd to have the ſtatute labour of the whole kingdom regulated by one general law, ſeeing that, in ſome diſtricts, from the nature of the ground, and ſcarcity of materials, the expence of repairing them, is more than double what it is in others. We would therefore recommend an alteration of the law in thoſe reſpects, that the tax ſhould be levied in an equal and juſt way, by a parochial or county rate upon all perſons, in direct proportion to the benefit they received from the roads; and that coaches, chaiſes, and ſaddle horſes, kept by landed gentlemen and others, ſhould pay, which are at preſent totally exempted. If this rate was made to riſe or fall according to the good or bad condition of the roads, we entertain the hopes, that the whole roads in the iſland would ſoon be in a comfortable ſtate of repair, and conſequently the facility and pleaſure of travelling, greatly increaſed.

Before finiſhing this ſection, we are called to notice the loſs and hardſhip ſuſtained by many roads, in conſequence of the mail coaches being ſuffered to travel over them, without paying any thing towards their ſupport. Theſe machines, from their great weight, and from the ſpeed with which they are driven, do amazing damage to the roads over which they paſs, and will ſoon either occaſion a bankruptcy in ſome diſtricts, or an increaſe in the rate of tolls. We do not pretend to be acquainted with the profits of the contractors employed, and perhaps common report magnifies them; but whatever they may be, there can be no valid reaſon offered, why a particular diſtrict or diſtricts ſhould be ſaddled with additional expences, upon account of a conveyance, in which the public at large are equally intereſted.

Another thing which deſerves to be noticed, is the low rate of turnpike duty paid by waggons furniſhed with broad wheels. We have already ſaid, that theſe waggons occaſion great damage to the roads; perhaps one of them does more hurt than twenty ſingle carts, and yet they have, in every turnpike act, been ſo far favoured by the legiſlature, as to be ſubjected only to half duty. Theſe vehicles, from the manner in which their wheels are ſhoed, and from the heavy loads put on them, preſs down the hardeſt laid road, and when dragged down a hill, they make a rutt or track ſomething like a plough furrow. We cannot but recommend, that full tolls ſhould be laid on broad wheel carriages, which would diſcourage the uſe of waggons; and their ſuppreſſion would not only be of immenſe benefit to the roads, but very advantageous to every perſon employed in the tranſportation of goods from one place to another.

SECT. 2.—*Canals.*

INLAND navigation or canals, fall next to be considered after public roads, and they are of real importance, by allowing the numerous and bulky articles manufactured in the district, to be transported from one place to another, at a less expence than they could be conveyed by the best repaired roads. In this point of view, independent of private advantage, Inland navigation cannot be too much recommended, wherever the nature of the country will admit of it, and where the trade of the neighbourhood is extensive enough to defray the charge.

We cannot speak with certainty respecting the extent of inland navigation in the West Riding, but it appeared to us, that the district was well supplied in this respect, and indeed the trade of the country could not otherwise be carried on to advantage.

SECT 3.—*Fairs and Markets.*

THE following is the most accurate account we could procure of the different fairs held in the West Riding, and of the articles offered for sale at them.

Aberford.—Last Wednesday in April, last Wednesday in May, last Wednesday in October, and Wednesday after St Luke, October 18th, for horses, horned cattle and sheep.

Adwalton.—January 26th, February 26th, Thursday in Easter week, Thursday fortnight after Easter, Thursday month after Easter, Whit. Thursday, and every Thursday fortnight after till Michelmas, for horses, cattle, pedlary, &c.

Boroughbridge.—April 27th, for horned cattle and sheep. June 22d, for horses, horned cattle and sheep, and hardware. October 23d, for ditto.

Barnsley.—Last Wednesday in February, preceding 28th; if Wednesday be the 28th, it is held Wednesday before, so that it can never be later than the 27th, or sooner than 21st, great fair for horned cattle and sheep. May 12th ditto. October 10th ditto.

Bawtry.—Holy Thursday, Old Martinmas, November 22d, for cattle and horses.

Bingley.—January 25th, for horned cattle. August 25th, 26th, 27th, for cattle, sheep, and linen.

Bradford.—March 14th, 15th, June 28th, 29th, 30th, for cattle and household furniture. December 20th, 21st, 22d, large fairs for hogs.

Bentham.—June 24th, for cattle.

Bradfield.—June 17th, December 9th, chiefly for swine.

Cawood.—May 12th, for cattle and wooden-ware.

Clapham.—St Mathew's, September 21st, for sheep.

Doncaster.—April 5th, August 5th, November 26th, and Monday before Old Candlemas Day, for horses, cattle, sheep, and pedlary.

Dewsbury.—Wednesday before May 12th, Wednesday before October 10th, for horned cattle and sheep.

Gargrave.—December 11th, for horned cattle and toys.

Gisburn.—Easter Monday, Monday fortnight after Easter, Monday month after Easter, Saturday after Monday month from Easter, for horned cattle. Monday 5 weeks after Easter, for pedlary. September 18th, 19th, for horned cattle and pedlary.

Halifax.—June 24th, for horses.

Helmsworth.—October 30th, for horned cattle.

Huddersfield.—May 24th, for lean horned cattle and horses.

Ingleton.—November 17th, for leather and oat meal.

Keighly.—May 8th, for horned cattle, brass, and pewter. November 8th, for horned cattle, brass, pewter, and pedlary.

Knaresborough.—Wednesday after January 24th, Wednesday after March the 12th, May 6th, Wednesday after August 12th, Monday after October 10th, December 13th, for horned cattle, horses, hogs, and sheep.

Lee, otherwise *Leegap.*—August 24th, September 17th, for horses and cheese.

Leeds.—July 10th, for horses and hardware. November 8th, for horned cattle, horses, and hardware.

Otley.—August 1th, November 15th, for horned cattle, and houshold goods.

Penniston.—Thursday before February 28th, the last Thursday in March, Thursday before Old May Day, and the Thursday after Old Michelmas Day, for horned cattle and horses.

Pontefract.—St Andrew's fair, on the first Saturday in December; twenty day fair, the first Saturday after the 20th day from Christmas; Candlemas fair, and first Saturday after February 13th; St Gile's fair, the first Saturday after September 12th; and all the other moveable fairs, viz. Palm Sunday, Low Sunday, and Trinity Sunday, to be held on the Saturday before each of these days respectively. The fortnight fairs will always be held on the Saturday next, after York fortnight fairs, as usual. The shew of horses formerly called Palm Sunday shew, will always for the future begin on the 5th of February.

Ripley.—August 25th, 26th, 27th, for sheep, horned cattle, and linen.

Ripon.—Thursday after January 24th, Thursday after March 21st, for horses, horned cattle, and leather. May 12th and 13th, for horses and sheep. First

Thurfday in June, horned cattle, horfes, leather, and fheep. Holy Thurfday, firft Thurfday after Auguft 22d, November 22d, for horfes and fheep.

Rotherham.—Whit-Monday, for horned cattle, and fheep. December 1ft, for cattle and horfes.

Sedbergh.—March 20th, October 29th, for horned cattle.

Selby.—Eafter Tuefday, June 22d, October 10th, for cattle, wool, linen, tin, and copper ware.

Settle.—Tuefday before Palm Sunday, Thurfday before Good Friday, and every other Friday till Whitfunday, for horned cattle. April 26th, for fheep. Auguft 18th, to 21ft, firft Tuefday after October 27th, for horned cattle, leather, wool, fheep, lambs, &c.

Sheffield.—Tuefday after Trinity Sunday, November 28th, for cattle and horfes.

Sherburn.—October 6th, for flax and horfes.

Slaidburn.—February 14th, April 15th, Auguft 1ft, October 20th, for cattle.

Snaith.—Laft Friday in April, Auguft 10th, for cattle, horfes, and pedlary. Firft Friday in September, for cattle and horfes.

Thorne.—Firft Monday, Tuefday, and Wednefday, after June 11th, and alfo the faid days after October 11th, for horned cattle, horfes, and pedlary.

Topcliff.—July 17th and 18th, for fheep, horned cattle, horfes, &c.

Wakefield.—July 4th and 5th, for horfes and hardware. November 11th and 12th, for horfes and horned cattle; if either of thefe days fall on a Sunday, the fair is held on the Saturday before. Note, July 5th, and November 12th, are pleafure fairs, toys, &c.

Wetherby.—Holy Thurfday, Auguft 5th, November 22d, for horfes, fheep, and hogs.

Whitgift.—July 22d, for pedlary.

The market towns in the West Riding are

Leeds,	Settle,	Wetherby,
Wakefield,	Ripley,	Sherborn,
Halifax,	Rippon,	Aberford,
Bradford,	Boroughbridge,	Cawood,
Huddersfield,	Aldborough,	Gisborn,
Sheffield,	Knaresborough,	Selley,
Doncaster,	Otley,	Tadcaster,
Pontefract,	Rotheram,	Bawtry,
Barnsley,	Snaith,	Tickhill.
Skipton,		

A very considerable corn market is held at Knaresborough, where dealers from the western parts of the Riding attend, and purchase from the farmers in that neighbourhood. A great part of this is resold at Skipton market in Craven, and carried still farther westward, where corn is scarce, and gives employment to a number of people who are concerned in this traffic.

It is under circumstances of this kind that public markets for grain can be considered as advantageous to the growers or purchasers of corn. The first cannot get his commodity disposed of at home, hence willingly goes a stage to meet his merchant; and the latter being sure to meet with a supply, attends upon market day, with his horses and carts, for conveying it to the place where he is to use it, or dispose of it again. By this mode, no time is lost, no unnecessary labour incurred; whereas, were all the grain in the kingdom to be sold in the public market, as some wild imaginations recently proposed, a great waste of both must necessarily happen.

Let us just suppose that such a law had been passed, and that the grain sold at Knaresborough was not to be drove to the west bounds of the Riding, but that it was wholly to be consumed in the neighbourhood of that place; and say, where would be the advantage arising from setting down the sacks in the market? It might happen, that a baker or maltster purchased the very wheat or barley which was grown by his next door neighbour, but which, in consequence of such a mistaken law, could not be sold without being first offered to sale in this public manner. Would not the trouble of driving it to market by the farmer, and of driving it back again by the baker or maltster, be just so much lost labour to them, without affording the smallest advantage, nay, rather occasioning a positive loss to the public consumer, upon whom every expence of this kind must necessarily fall.

SECT. 4.—*Manufactures.*

THE manufactures of the West Riding are numerous and valuable, and comprehend broad and narrow cloths of all qualities, shalloons, calimancoes, flannels, and every branch of woolen goods. The manufacture of these articles is carried on at Leeds, Wakefield, Bradford, Halifax, and Huddersfield, and in the country adjoining to these places, to an astonishing extent. The whole wool of the district is not only wrought up in these manufactures, but immense quantities are also purchased in the conterminous counties for the same purpose.

While the people in the heart of the district are thus employed in manufacturing woolen goods, those of the southern parts are engaged in carrying on manufactures no less valuable, and fully as important. At Sheffield and its neighbourhood, every kind of cutlery and plated goods are manufactured; and so eminent are the artizans in their different professions, that no other place is able to compete with them in the manufacturing of these articles. Sheffield has been a staple place for knives for more than three hundred years, as may be inferred from Chaucer, who says in his poems,

"A Sheffield whittle bore he in his hose."

And Leland observes, that great numbers of smiths and cutlers lived in those parts when he wrote, which was in the reign of Henry VIII.

Rotherham, in the neighbourhood of Sheffield, is a place famous for iron works, similar to those carried on at Carron, in Scotland. We here saw a part of the stupendous iron bridge lately erected over the river Ware, at Sunderland, which was executed by the Messrs Walkers, proprietors of these works. The merit and ingenuity of these gentlemen, deserve every mark of public encouragement.

The establishment of manufactures in the West Riding has been the principal cause of its present wealth. It is difficult to ascertain the period when they were first introduced, but there is reason to suppose, it was about the beginning of the fifteenth century. Camden, in his Britannia, fixes the introduction of manufactures to have been during the reigns of Henry VIII and Edward VI. This æra may, however, be suspected; for there is a copy of a court-roll, as we were informed, still extant, dated at the court of the Prior of Lewes, held at Halifax on the Thursday after the Feast of St Thomas, 2d Henry V, 1414, wherein Richard de Sunderland, and Joan his wife, surrender into the hands of the lord of the manor, an inclosure at Halifax, called the *Tenter Croft*; which is a strong presumption that manufactures were carried on there before that period.

The country chosen for carrying on these manufactures is admirably adapted to that purpose. The raw materials are abundant on every hand; and coals, which are indifpenfably neceffary, are plentiful and cheap. The ground in the vicinity of the manufacturing towns has in general been originally barren, and in many parts little better than wafte; but from the great increafe of population, and the additional quantity of manure occafioned by the manufactures, the foil is now equal in value to that of places originally more fertile.

It appears to us, that manufactures have had a fenfible effect in promoting agriculture in this diftrict. By them a ready market is afforded for every article of provifions that can be raifed, without which agriculture muft always be feeble and languid. They have, no doubt, raifed the rate of wages bonfiderably: this always follows of courfe, where trade profpers, and is a fure fign of wealth; but they have at the fame time raifed the value of the produce of land, which much more than enables the farmer to pay the increafed rate of wages.

From all the enquiries we could make, we did not find that the effects of manufactures were detrimental to agriculture, by rendering hands fcarce for carrying it on. In harveft the manufacturers generally leave their looms, and affift in reaping the crop. We did not hear of any feafon when hands could not be found fufficient to anfwer the demand, except in 1792, at which time the manufacturers had orders to an uncommon extent. Even then, this fcarcity was no further felt in the Weft Riding than by a great rife of wages; although we were informed, that in the Eaft Riding a very heavy lofs was fuftained.

A confiderable portion of the land is occupied by perfons whofe chief dependence is upon manufactures. We are not, in this cafe, to expect the fame attention to the minutiæ of farming, as from thofe who make it their fole occupation. Their minds and capitals are generally fixed upon their own bufinefs, and land is folely farmed by them as a matter of convenience or amusement. In the vicinity of the manufacturing towns, great numbers of milch cows are kept, and there is a constant demand, not only in those places, but over the whole Riding, for milk, and the articles of cheese and butter, which are produced from it.

We have already said, that the soil in the manufacturing district has been originally of the most barren sort, and in many parts little better than waste. It may be remarked on this subject, that in those counties where the soil is proper for carrying on agriculture, the disposition of the people is always inclined to rural affairs; while, in other parts, where the soil is sterile and unproductive, the genius of the people is turned to manufactures and trade. This remark will with much truth apply to the greatest part of Britain, and is a demonstration, that the bounties of nature are dispensed in an equitable manner. While the inhabitants of the favoured soil raise corn for the support of the community, those who are not blessed in this way, manufacture goods for the comfort and convenience of the happy agriculturist, and in this manner both equally promote the public good.

We are furnished with two papers, containing valuable observations on manufacturers residing in the country, and occasionally employed in cultivating the soil, which we with pleasure insert.

Hints, Respecting Manufacturers residing in the Country, who are occasionally employed in Cultivating the Soil.

"The few observations which the writer is able to furnish upon this chapter, will be confined to such manufacturers as are employed in the making of woollen and worsted goods, exposed to sale in the different market-towns of Leeds, Wakefield, Huddersfield, Halifax, and

Bradford," in the West Riding of the county of York, and which consist of broad and narrow cloths, shalloons, callimancoes, and the various worsted articles, which the industry and ingenuity of the persons employed have diversified and improved; and in considering the question, it is the writer's opinion, that those manufacturers have many advantages by residing in the country. For,

" *1st*, They enjoy a more uncontaminated air, which, as the employment of the clothiers is not the most cleanly, will conduce to their health.

" *2dly*, The country affords them a more open exposure of their manufacture to the sun, which is necessary in different stages of their work.

" *3dly*, In general, the villages where the manufacturers are resident, are nearer to, and more cheaply supplied with coals; an article, not only necessary to the comfort of their families, but also to enable them to carry on their trade.

" *4thly*, Another advantage attending a country residence, is the many springs of good wholesome water for the supply of their families and their dye houses; for it is to be observed, that every clothier dyes his own wool, unless colours are required of uncommon brilliancy.

" *5thly*, Another advantage is, that by being thus disposed in villages, the manufacturers are nearer to the fulling mills, with which the different rivers are occupied; and it is this dispersion which has occasioned so many fulling mills to be erected, to the great advantage of the owners of the different falls upon the rivers, which otherwise would have been almost useless.

" *6thly*, The manufacturer of cloth in particular, requires roomy buildings, which are obtained upon much lower rents in the country than in towns.

" *7thly*, From the bulkiness of the raw materials, and upon various other accounts, a horse is almost necessary to enable a clothier to carry on his trade; and as land at a distance from large towns, is cheaper generally than near them, the manufacturer in the country can better keep so useful an animal.

"*8thly*, To do this, and also to maintain a cow, which is one of the first comforts and chief supports of the infant part of his family, the country affords him a much cheaper, and better opportunity; and as both hay and straw are wanted for the animals, the manufacturer, partly of necessity, occasionally becomes employed in the cultivation of the soil, and it is no uncommon thing to see, in a manufacturing farm, which ought not to exceed (and seldom does) 16 acres, great attention, judgment, and spirit, in cultivation. Certain it is, that by manufacturers residing in the country, and occasionally employing themselves in cultivating the soil, the barren commons of these parts, a great many whereof have been lately inclosed and divided, have been made productive to a degree, which no regular farmer could have made it their interest to have attempted. By thus becoming the cultivator of land, the manufacturer is enabled to raise poultry, and keep a pig, and, accustomed to cut his own corn, he becomes acquainted with the sickle, which he is called forth frequently to use in the harvest of the country, where more corn is grown, and where there are fewer hands to get it in.

"*Lastly*, By living in the country there is less temptation to vice; and by occupying a small parcel of land, a life of labour is deversified, and consequently relieved."

Observations Respecting Manufacturers being partly Employed in the Cultivation of the Soil.

With respect to the manufacturers residing in Yorkshire, they seldom are farmers of land, beyond the conveniencies and exigencies of their trade. A home-stead, a sufficient quantity of meadow and of pasture for the support of a horse and a cow, with now and then a corn field, form, with few exceptions, the extent of their speculations in agriculture. The necessity of their possessing the two first, operates so as to occasion their being obliged to give a high price for these accommodations, compared with that of land in such neighbouring townships, as are not inhabited by manufacturers.

"The high wages which a working manufacturer can earn, exceed so much the usual prices of agricultural labour, that the master manufacturer seldom keeps any other than of the former description on his farm. In short, the manufacturers in the West Riding of Yorkshire, have little, if any pretensions, to the character of farmers. The speculations, the interruptions, inseparable from trade, call for all his capital; and (unless in some particular cases, where a manufacturer happens to have land by inheritance, or an advantageous lease), his time, circumstances, and interest, conspire to prevent him from following up both professions at one and the same time.

"In Yorkshire, the master manufacturers reside in villages, and bring their goods to the several halls of Leeds, Wakefield, Huddersfield, Halifax, and Bradford, for sale. In Lancashire, the woollen trade is carried on differently: The master manufacturers are comparatively very few in number; these vend their own goods to the merchants and shopkeepers; and the having a farm in their own hands is not unfrequent, nor incompatible with their other professions. But here again the enhanced price of land in all manufacturing districts, admitting the soil and situation were suitable, is adverse to their growing much corn."

SECT. 5.—*Corn Laws.*

IT would perhaps be improper, in a local survey, to enter upon a regular examination of the corn laws; but as every farmer in the kingdom is less or more interested in such an enquiry, we cannot pass over the subject altogether.

The old corn laws of Britain were enacted upon the supposition, that more corn was raised in the island than the consumption of the inhabitants required, and that to procure a market for what remained on hand, it was expedient to grant a bounty to the exporter, so as he might be able to meet the foreign merchant upon equal terms. We are not here to enter upon the question, whether this bounty was meant as a reward to the landed interest for supporting the Revolution, as has been often alleged; but it certainly contributed to keep up the prices of produce, by enabling the British farmer to compete with his foreign brethren, who raised their grain at less expence, who paid less rents, and who were not subjected to such heavy taxes; and so long as Britain raised a greater quantity of grain than was necessary for supplying the internal demand, the law of 1689 must be considered as founded in policy and wisdom.

From the beginning of this century, to the year 1756, the corn laws were allowed to operate without any suspension; but the crop of the above year being rather defective, an act of Parliament was passed, whereby exportation was stopped during the year 1757. In 1766, upon an application from the Lord Mayor of London, the Privy Council assumed the power of issuing a proclamation for stopping exportation, which was emphatically called by the late Earl Mansfield, "the forty days "tyranny;" and since the year 1773, the corn laws, like Proteus, have aſſumed ſo many various ſhapes, no regular ſyſtem being adhered to, as to throw the whole trade into confuſion and diſorder.

It has been argued, that the bounty rendered corn cheaper at home, by encouraging tillage, and that to its operations, the great improvements in Britain muſt be attributed. Upon the firſt point we are rather ſceptical, for an increaſed demand for any article of trade, certainly ſerves to raiſe the market; and although that increaſe may be beneficial to the ſeller, it never can enable the purchaſer to buy ſo low as if there were fewer competitors. Again, if the bounty has increaſed rents, the farmer has thereby paid away all the advance he received in conſequence of his acceſs to foreign markets, and tillage has received no greater encouragement than it would have experienced, had no ſuch laws been paſſed.

Whatever ſhould be the reſult of theſe arguments, when applied to the times when Britain produced more corn than was neceſſary for ſupplying the home conſumption, they do not fall to be taken into conſideration, when our conſumption is undoubtedly greater than what our produce can ſupply.

But, ſay the gentlemen who ſupport the old ſyſtem, "That very decreaſe of produce you are ſpeaking of proceeds entirely from the alteration of ſyſtem; reſtore the old corn laws, and grain will be both plentier and cheaper. No encouragement is given to agriculture, but the intereſt of the manufacturer is alone regarded. Tillage is diſcouraged, and the farmer is obliged to throw his lands into graſs, which renders corn ſcarce." Theſe things have often been urged, and we ſhall juſt ſay a few words in anſwer.

If prices are to be conſidered as a criterion for judging of the encouragement given to any trade, certainly the tillage farmer has for ſeveral years back received ſufficient ſupport; but, have theſe prices been of advantage to the profeſſion in general? have they not ſerved to raiſe rents to the moſt extraordinary pitch, and alſo increaſed the value of all kinds of labour in a ſimilar manner? The farmer rather ſtands in need of protecting laws to ſave him from ruin, as, from the burthens accumulated upon him, it is next to impoſſible he can ſell his commodities at the ſame rates they were ſold at in former times.

Now, if a renewal of the old ſyſtem of corn laws were to make markets cheaper, the ruin of the farmer would be haſtened, inſtead of encouragement being afforded to the culture of grain. We are advocates for theſe laws in one inſtance, becauſe their operation was ſteady and regular; whereas, the innovations introduced ſince 1773, and principally in 1791, have rendered the corn trade like a lottery, and have ſet the diſcernment of the wiſeſt at defiance.

But, have theſe modern laws injured tillage, by cauſing greater quantities of land to be thrown into the grazing huſbandry? No, they have not; for great as the quantity of paſture and meadow is in Britain, ſtill it is below the demand, which is confirmed by the aſtoniſhing prices of all ſorts of ſtock. Luxury has of late increaſed with ſuch rapidity, that a far greater number of acres are now required to ſupport the ſame number of inhabitants than formerly. Whenever there is more graſs than the demand requires, the diſeaſe will inſtantly work its own cure. When cattle and ſheep cannot be ſold, the paſtures will be broke up, for corn is an article that will ſooner find a market than butcher meat.

To ſum up what we have ſaid, it appears to us, it would be of public advantage, that the corn laws were regulated upon ſome permanent plan; and, that under exiſting circumſtances, there is no cauſe for a bounty being granted upon exportation, which even in former times, occasioned innumerable frauds. Perhaps, under the present burthens, there is a necessity for raising the importation prices; for, laying rent out of the question, the value of labour and public burthens, oblige the farmer to raise grain at the increased expence of twenty-five per cent. above what he could do it at, twenty years ago. Now, if the importation price of wheat at that time was 48s, it appears the farmer has a right to its being now advanced to 60s, independent of the rise of rent, which, during the same period, has increased in a far greater degree than either labour or public burthens.

Before we finish this article, we are called upon to notice the temporary expedients of the day, which have been adopted respecting the corn laws since 1791. We cannot set the errors of them in a stronger light, than by stating, that many farmers who have taken land upon the faith of the law 1791, are exposed thereby to ruin and destruction. They reasoned in this manner:—" The Lords of his Majesty's Privy Council have given it as their opinion, that Britain does not, upon an average of years, grow corn sufficient for the consumption of its inhabitants; and the Legislature have passed a law, declaring that wheat cannot be imported, duty free, before the home prices are 54s. per quarter. The inference therefore is, that 54s. must be nearly the medium price, so long as the state of husbandry remains upon its present footing, and this law shall continue in force." The farmer therefore makes his calculations accordingly, for some data or other he must assume; but, to his surprize, when the state of his crop appeared to warrant prices above an average, a suspending power starts up, a power unknown in this country since the Revolution, which reduces the laws he depended upon into a non-entity, and allows the country to be inundated with foreign grain, while prices are much below what the law declared ſhould be the importation rate. This is the real defect in our exiſting laws, and they ought inſtantly to be corrected in ſuch a way, as not to leave their execution to depend upon the caprice or timidity of any man or ſet of men whatever.

Sect. 6.—*Poor and Population.*

If we might venture to form a calculation reſpecting the annual expence of the poor in this diſtrict, we would ſay, it was equal to one fifth of the rental; but as, notwithſtanding our utmoſt reſearches, we cannot give a correct ſtatement of this expence, we ſhall only ſay, that it has of late greatly increaſed, eſpecially in the neighbourhood of manufacturing towns, and if trade declines, a ſtill greater increaſe may be expected.

We are inclined to think, the funds for the ſupport of the poor are managed in a judicious way; at leaſt, notwithſtanding our enquiries, we have heard of nothing to the contrary, but rather in favour of thoſe to whoſe charge they are entruſted. Removals are complained of as being attended, in many caſes, with a greater expence than what would be required for ſupporting the pauper.

The population of the Weſt Riding is great, and very probably upon the increaſe. From the beſt accounts we could obtain, it is calculated at upwards of four hundred thouſand: but, where an actual numeration is not made, the beſt accounts may be defective. Owing to the populous manufacturing towns, it muſt at any rate be great, and the number of men liable to ſerve in the militia, confirm the ſtatement we have given.

NOTE on Chap. 15.

(*a*) The owner of a four wheeled chaise ought to pay three or four times as much to the roads as that of a waggon. He is much more interested in a good road than the farmers, and generally pays little or nothing.

W. P.

Remarks on the above Note.—If the rule for taxation was the ability of the person who was liable, the owner of a chaise ought undoubtedly to pay much higher to the support of the roads than the humble owners of waggons and carts, who derive a living from hiring out these vehicles; but whatever arguments may be used for continuing toll money at the rates now payable, none can be urged in favour of coaches, chaises, and pleasure horses, being totally exempted from statute labour upon the bye-roads. Throwing the repair of the latter upon the farmers, may justly be considered as a remnant of feudal servitude, now proper to be removed. *R. B.*

CHAPTER XVI.

MISCELLANEOUS OBSERVATIONS.

Sect. 1.—*Agricultural Societies.*

IN our progreſs through the Weſt Riding, we could not learn, after the minuteſt inquiry, that a ſingle ſociety ſubſiſted for the improvement of agriculture. We heard of three that were formerly eſtabliſhed for that uſeful purpoſe, viz. at Sheffield, Bautry, and Doncaſter, but theſe for ſome time paſt have been diſcontinued.

As improvements in agriculture very often locally take place, and are ſlow in travelling from one part of a country to another, we ſhould eſteem the inſtitution of ſocieties, upon proper principles, an excellent method for diſſeminating knowledge in this ſcience; if theſe ſocieties were to correſpond with one another, every new improvement, either in cultivation, ſtock, or huſbandry utenſiles, that was deviſed in one part of the country, would be immediately known in its moſt diſtant parts. For want of theſe means of communication at preſent, the great body of farmers are almoſt as ignorant of what their brethren in other counties are doing, as if they lived in a foreign land (*a*).

In conſtituting agricultural ſocieties, we are far from recommending an intermixture of proprietors and farmers together (*b*). It is abſolutely neceſſary, for many obvious reaſons, they ſhould be ſeparate. Without dwelling upon these, it may only be said, that, in presence of his landlord, the farmer is too ready to be diffident, and will not propose his opinions in that free and unrestrained manner he would do, if only amongst the company of his brethren and equals (*c*). We heard of the Sheffield society, where gentlemen, clergy, and farmers, met promiscuously; the consequence of which was, that the latter were in a manner prohibited from mentioning improvements, in case they should be a watch-word for the one to increase the rent, and the other to raise the rate of tithes.

Since writing the above we learn, that an agricultural society has been lately established in the West Riding, under the patronage of the Right Honourable Lord Hawke; but as we have not been favoured with the plan or bye laws of the society, we cannot enter upon particulars.

SECT. 2.—*Weights and Measures.*

FEW subjects are of greater importance to the agriculturist, than a reduction of the weights and measures of the kingdom to settled standards, as their present disordered state is productive of innumerable evils, both to the seller and the buyer of every home raised commodity. Indeed, it is only in the sale of our own produce that the diversity is felt, as every foreign article is uniformly sold by fixed and well known standards: But, notwithstanding this business has repeatedly occupied the attention of the Legislature, the confusion, which has for ages prevailed, is still suffered to remain.

A bill was introduced into last Parliament, which promised to do away every evil sustained from the discrepancy of meafures, but as that bill has fome way or other been fince neglected, although at the time it met with general approbation, we think it a duty incumbent upon us to ftate a few of the pernicious confequences arifing from the confufion amongft the meafures of capacity, and that it would be of great public advantage, if all corn was fold by weight, as propofed by the above bill.

1ft, The grower of corn is thereby expofed to various impofitions in the fale of his produce. He firft meafures the corn at home, and when it is delivered, the purchafer, if he pleafes, may infift upon having it re-meafured by the ftandard of the place; and, if the meafure is in the leaft defective, a deduction muft be made. If thofe ftandards were exact, no complaint could juftly be made, on account of this deduction; but when it is confidered, that almoft every one of the municipal ftandards are larger than they ought to be, it is more than prefumable that deficiencies are often demanded, when the full legal quantity is actually delivered. Nor is it an eafy matter to procure redrefs, the exact fize, or cubical contents of the ftandard meafures, being but imperfectly known among farmers; and even if they were fufficiently known by them, few municipal officers would be found willing to lend an ear to their complaints. So far from that, the municipalities of the kingdom have, in a great meafure, been the authors of the prefent confufion, and they are interefted in keeping it up.

2dly, The diverfity of meafures is injurious to the confumer of bread, becaufe the affize of that neceffary article is thereby fixed higher than it ought to be, or would be, if uniformity prevailed. So long as meafures remain unequal, it is impoffible to regulate the price of bread upon any thing like juft principles. No; that can only be done where meafures are uniform, or where corn is fold by one fixed rule.

3dly, The diverſity of meaſures is injurious to the public at large. The corn trade of Britain is of great and general importance, and the import and export of that neceſſary article, affects the intereſt of a greater number of people than any other meaſure of political œconomy. By the exiſting laws, the ports for importation and exportation are opened and ſhut according to the liſts of average prices, returned by the different counties, to the corn inſpector at London. When the amazing differences among the cuſtomary meaſures are conſidered, it will be found totally impoſſible to reduce them correctly to the ſize of the Wincheſter buſhel. Hence the groſſeſt errors are to be found in thoſe returns; which any perſon muſt be ſatisfied of, by examining the general averages of the diſtricts, into which the kingdom is divided; nor can it be otherwiſe, ſo long as a diverſity of meaſures continues. The errors of theſe averages equally affect both the grower and the conſumer of corn. If the ports are opened earlier than they would have been, had the average been fairly aſcertained, the farmer, who probably took his farm upon the faith of foreign corn being excluded, before prices reached a certain height, is neceſſarily injured; while, *vice verſa*, if the ports are kept longer ſhut, than a fair average would have warranted, the conſumer has an equal right to complain.

But why need we attempt to prove the baneful conſequences attending the diverſity of meaſures, ſeeing they have been acknowledged in all ages, and have occaſioned numerous laws being paſſed to procure uniformity. As theſe laws, from ſeveral cauſes, have failed to produce the intended effect, the concluſion muſt be, that ſome other mode of ſelling corn muſt be reſorted to, before we can enjoy the beneficial conſequences of uniformity. It appears, the gentlemen who framed the bill above mentioned, viewed the bufinefs in this light, and therefore propofed in future, that grain of all kinds fhould only be fold by weight.

It is unneceffary to inquire, at what period meafures of capacity were introduced into this country? But there certainly was a time when corn, as well as every other commodity, was bought and fold by barter. Let us, therefore, return to firft principles, which is always the fureft way for rectifying abufes. Let us fuppofe we had no eftablifhed meafure of capacity, for felling the produce of the foil, but that every part of it was exchanged for what it would bring. Let us alfo fuppofe, that the country poffeffed a ftone, called a pound, and that its weight and fize were exactly afcertained. Under thefe fuppofitions, could any thing be more natural and fair, than for the perfon who poffeffed corn, to fay to him with whom he had been in ufe of bartering that article, " We have hitherto been dealing upon loofe principles, I have given you the produce of my land, but am ignorant of the quantity you received, and of the value of the article I got in return. I will therefore give you corn according to the weight of this ftone, for fo much money, and our dealings will not be expofed to the uncertainty I am complaining of." A propofition of this nature, fo candid and equitable, it is prefumed, would be inftantly accepted, and would, from that time, be the rule of their future tranfactions.

Upon principles fomething fimilar to the above, has the bill been framed, which was introduced into the laft parliament, for regulating the fale of corn by weight, and which, if paffed into a law, promifes to be a radical cure for the abufes, which, from the lapfe of time, or other caufes, have crept in amongft the meafures of capacity. Independent of correcting thefe abufes, the felling of corn by weight is the moft equitable way of difpofing of it. The feller muft receive a price in direct proportion to the quality of his grain, unlefs he foolifhly fells his fack, or bag, or hundred weight, at a lower price than it is worth. Good grain, when properly dreffed, will infallibly draw its fair market value; which is not always the cafe when fold by a meafure of capacity, and the fineffes practifed in filling and rolling the meafure, will be effectually prevented.

It has often been remarked, that neither good land nor good corn draw a price in the market, proportional to their intrinfic value, when compared to inferior forts. This remark, fo far as it applies to corn, appears to us to be juft; felling it by weight, will therefore, in a great meafure, remove the objection, as hufks and fcales, although they fill the bufhel, full as well, if not better than found corn, will go fhort way in bringing down the arm of the beam. We are fenfible inferior grain has more refufe, than is to be found amongft the like quantity of good grain; but this refufe cannot affect the purchafer fo much when weighed, as when delivered by a meafure of capacity; at any rate, the fkill of the purchafer, muft in this, as well as in every other tranfaction, be his guide in fixing the price.

It may probably be urged againft our argument, that the value of grain cannot be afcertained by its weight; that a hundred pounds of wheat, produced upon a good foil, and in a favourable climate, will yield more flour, and of a fuperior quality, than a hundred pounds of wheat raifed on a more unfavourable foil, and in a worfe climate. To this we anfwer, that if the bill fixed the price of a fack of wheat, as it does the number of pounds the fack fhall contain, the objection would be well founded: but this the bill does not interefere with; it leaves the price or value of the fack to be fettled between the feller and the buyer, and only fecures the latter in the full quantity he understands he is to receive when he purchases a sack. At the same time, no person of common experience will deny, that there is a far greater difference betwixt the produce, in flour, of a bushel of good wheat, and a bushel of an inferior quality, than what will be betwixt 280 lbs. *i. e.* a sack, of two kinds under similar circumstances. If 280 lbs. of each kind were measured, the one would be nearly a bushel more in measure than the other.

Another advantage that may arise from selling corn by weight is, that it should induce every farmer, both from principles of honour and interest, to dress his grain in a sufficient manner, and to keep the lightest of it at home for domestic consumption. It is plain, that so long as measures of capacity are used, this will not be studied. Selling it by weight will therefore remove every temptation to deliver corn, unless in its most perfect state.

A law for selling corn by weight, will at once annihilate the anarchy and confusion which the discrepancy of measures has introduced into the corn trade. Corn is the staff of life, and the cultivators of the ground may be considered as the first and most valuable of all our manufacturers. The importance of the corn trade claims every mark of legislative attention; and sound policy and true wisdom, call for countenance and protection to those employed in this, the most valuable as well as the most necessary of human arts.

It would be none of the least advantages of the proposed bill, that it prevents all those slight-of-hand practices used in filling and rolling a measure of capacity, which, under the management of a clever hand, are equal to *one per cent.* So long as measures are used, it is not to be doubted, that every person will endeavour to fill them as dexterously as possible; and, for doing so, no blame can be incurred. But weighing of corn puts every one upon an equal footing, and will also be the means of preventing thofe numberlefs difputes which continually happen in every market about the fize of corn meafures.

We have heard a few objections againft felling corn by weight, fuch as, "that the beam employed in weighing the grain, may have a fhort arm;" "that the grain might be damped with water;" "that duft might be mixed with the water ufed in damping the corn;" and, "that weights would lofe by ruft, and turn lighter every day." Thefe objections we confider as groundlefs and infignificant; but, in cafe they fhould in any manner contribute to prevent fuch a beneficial meafure from being paffed into a law, we fhall confider them feperately.

1. *The beam may have a fhort arm.* If this is meant that the beam kept by the farmer may be deficient, the remedy is apparent, as the purchafer has only to go to the public fcales. If it is meant that the public fcales may be in that defective ftate, we would afk what temptation could induce the magiftrate to commit fuch a fraud? *Query*, whether is it eafier to have a beam with a fhort arm, or to keep a fmall meafure? From the general difcrepancy of meafures, no obloquy attends the poffeffor of the latter, but difgrace and infamy would attend the man who was guilty of the former fraud.

2. *Corn may be damped.* If this fraud is practicable, we are at a lofs to difcover what advantage would accrue to the feller from committing it. The feller of raw corn can never expect the fame price for a fack, as he who prefents it in a fufficient condition, for what he gained by increafing the weight, would be much more than loft by deficiency of the price. If it is meant that the corn may be fold by fample, and damped betwixt and the day of delivery, the anfwer is obvious, the buyer is not obliged to receive it; befides, the damping of corn

increases the bulk much more than it does the weight, which is an argument in favour of the bill.

3. *Dust may be mixed among the water used for damping the corn.* If a fraud of this nature was practicable, we should suppose it furnished an additional reason for passing the proposed bill, as the dust would swell the corn more than it would augment the weight; at any rate, if corn is damped or dusted, will not the eye and the hand of the purchaser direct him to find it out? We have often noticed great pains used to dry corn, so as to fit it for a market, the seller being conscious it was the only method by which he could dispose of it to advantage. Now, if this is thought necessary, when corn is sold by a measure of capacity, we may be certain it will be still more attended to when corn is sold by weight. Whatever is gained in quantity from selling raw corn is more than lost by the lowness of the price, independent of the trouble accompanying the sale of bad grain.

4. *But weights will lose by rust, and turn lighter every day.* Waving every previous argument which might be drawn from the established practice of this and other nations, of selling every thing by weight which can be ascertained by that method, we shall only notice, that the weights here meant, must be those kept by the farmer, as the objection was followed by remarking, that "they will generally be kept on the earthen floor of a barn." Whether keeping them in a barn will occasion a deficiency, we will not waste time in inquiring; for, so long as the public standards remain unimpaired, (which certainly will not be kept in a barn) any waste from rust, or other accidents, is easily remedied.

Having answered these objections, which were pretty generally circulated when the bill was under consideration, it only remains for us to add, that few objects deserve the attention of the Legislature more, than the regulation of the weights and meaſures, by which the produce of our ſoil is daily bought and ſold; and we truſt the buſineſs will ſoon be taken up in that ſerious and effective manner which its importance deſerves.

NOTES on Chap. 16.

(*a*) Agricultural ſocieties, upon a proper plan, might be highly beneficial to the community; but this cannot be accompliſhed, unleſs leaſes are granted, and tithes aboliſhed. A prudent man, who is tenant at will, or encumbered with tithes, is debarred from communicating his improvements, for obvious reaſons.

A Farmer.

(*b*) Societies of gentlemen might be of ſervice for the raiſing of ſubſcriptions, for the encouragement of improvements, in the management of land; but, if leaſes were granted, it would be unneceſſary, as the farmer's own intereſt would be a ſufficient inducement for his exertions, without the fear of danger from his communications. I would recommend thoſe ſocieties to conſiſt of farmers only; for gentlemen in this part of the country, know very little about the matter.

In ſome of the ſouthern counties, there are gentlemen of large properties, who have ſet laudable examples in the improvement of ſtock, &c. There are many of our gentlemen who know no more of thoſe animals, than thoſe know of them: In regard to proportion, a ram is a ram, and a bull is a bull; that is all they know or care. If gentlemen would truſt the management of their ſtock to judges, and let the uſe of the beſt bulls and rams be introduced amongſt their tenants, *gratis*, they would ſoon be convinced; but many of them will not purchaſe or hire, at *any rate*, what they don't underſtand. *A Yorkſhire Farmer.*

(*c*) There appears much plauſability in this reaſoning, and I am inclined to think, it will ſoon appear applicable to our Eſſex agricultural ſociety. *Anonymous.*

CHAPTER XVII.

OBSTACLES TO IMPROVEMENT.

HAVING detailed the prefent ftate of agriculture in the Weft Riding, and mentioned feveral obftacles to improvement, we now proceed to bring forward thefe obftacles in a regular manner: And, while we ftate our fentiments on thefe important matters with freedom, we truft that no partiality or prejudice fhall influence us to fwerve from a faithful difcharge of the truft committed to us upon this occafion.

When we entered upon the bufinefs of furveying the hufbandry of the Weft Riding, we were totally unacquainted with the practices and cuftoms of the diftrict; fo of courfe our minds might be fuppofed free of every kind of prejudice, when thefe were explained to us. We viewed a country bleffed with many local advantages; the foil in general much fuperior to our own; the climate comparatively good; markets for all kinds of produce quick and regular; rent on the whole lower than in other parts, not under fuch favourable circumftances; and yet, notwithftanding all thefe encouragements, the fituation of the farmer could not be confidered as comfortable, nor the practice of hufbandry fo perfect and correct as might have been reafonably expected. This led us to inveftigate the ftate of the country with minute attention; and the refult of our inquiries was, that hufbandry was not only retarded by feveral improper political regulations, but alfo by the nature of the connection which commonly subsisted betwixt proprietors and their tenants.

Under the first head, we beg leave to state the present situation of a considerable part of the Riding, occupied as common field, and of much larger tracts lying in a state of absolute waste. From the want of a general bill, these grounds cannot be divided, or held in severalty, without the proprietors incurring a vast expence by applications to the Legislature, which, in many cases, from the obstinacy and caprice of individuals, is not even practicable. We account it as demonstrable as any proposition in Euclid, that no real improvement can take place on the common fields and wastes, without a a previous division; and it is nearly as certain, that without a general law being passed at once for the whole kingdom, their division, according to the present system, will never be accomplished. We cannot display the difficulties which stand in the road of the proprietors of common fields and waste lands in a more pointed way, than what is done by the following petition to a Lord of the Manor in the West Riding, a copy of which was lately transmitted to us.

The Humble Petition of the Freeholders, Owners of Common Rights, and Occupiers of Lands and Messuages within the Township of ——.

Sheweth,

"That your petitioners, approach you with the profoundest respect, and cherishing a confidence in your attention to the public good, beg leave to signify to you the very heavy inconveniences your resiants of —— have long laboured under, from the want of inclosed ground in their neighbourhood; and at the same time to pray, that you will be pleased to join your petitioners in an application to Parliament, for an act to empower them to divide and inclose the several open fields and waſte grounds within the ſaid townſhip, as ſoon as may be.

" That the whole extent of the lands and grounds within the townſhip of ———, excluſive of the parks and grounds belonging to ——— ———, is computed to conſiſt of about two thouſand five hundred acres; and that not more than one-ſeventh part of theſe lands and grounds is at preſent incloſed; ſix-ſevenths of the ſame being open fields, commons and waſte grounds.

" That the value of the incloſed part, on account of the ſcarcity of incloſed land, is at preſent from one pound to two pounds per acre, and the value of the field land, in its preſent ſtate, no more than ſeven or eight ſhillings per acre; though, when incloſed, and duly improved upon the modern plan of cultivation, it would be equal in value to the preſent incloſures.

" That the neat value of the produce of one acre of incloſed ground, as at preſent cultivated, is at leaſt three times that of one acre of ground in the open fields, of the ſame quality with the former, taking the average value of three ſucceſſive crops: the reaſon of which diſproportion in ſuch values, is founded entirely in the different modes of cultivation; the owner of an incloſure being at liberty to adopt the moſt approved and profitable plan of cultivation, while the owner of the common field is compelled to follow the old one, of reaping two crops of corn the two ſucceſſive years after a fallow, and from which it is not in his power to deviate.

" That the waſte uninclosed ground, commonly called ———, conſiſting of at leaſt one thouſand acres, is in its preſent ſtate of very trifling value to the inhabitants of ———; and, if incloſed, would upon an average be worth at leaſt fifteen ſhillings per acre.

" That the townſhip of ——— is very favourably

situated for the improvement of land, on account of the great roads which intersect it.

"That the inclosure of the said grounds would greatly contribute to the breeding of sheep, and the growth of wool, as the land in the neighbourhood of ———, is of a kind most favourable to that mode of husbandry, which involves the keeping of sheep as its most profitable branch. The alternate and successive growth of corn, grass, clover, trefoil, turnips, &c. is very well known by the experienced farmer, to be the best and most profitable method of conducting the cultivation of a light, thin soil, such as that of ——— Hence, in all probability, a complete and regular plan of sheep keeping would in due time take place, in consequence of an inclosure. Plenty of pasturage would arise in summer from the white clovers and trefoils, and fodder in winter from the turnips and hay. An instance in point may be seen in the neighbouring town of Rigton; where considerable quantities of sheep are bred and kept: the land of which township is all inclosed, and exactly of the same kind with that of ———.

"That the farmers of ——— do not at present breed any sheep, from the want of inclosed land to produce fodder for the winter: nor do they find advantage in the few which some of them at present summer upon the common; for, in the first place, they are obliged to purchase their sheep from neighbouring villages or distant fairs in the spring; and, in the next place, they are under the like necessity of selling them off before winter to neighbouring farmers, as they are not in possession of sufficient inclosures to produce turnips and other winter fodder, to fatten them for the best market. The decline of sheep keeping here in general, and the circumstance, that many of the principal farmers, who formerly kept great numbers of sheep upon the common, have now discontinued the practice, attest, beyond contradiction, the inconsiderable advantage attending that branch of husbandry, in the present state of the lands of ——.

" That small as the profits are at present accruing to the town of ——, from the pasturage of cattle upon the Moor, they will undoubtedly very soon be smaller. For several of the owners of common rights upon —— Moor, induced by the certainty of making something of such rights, are beginning a practice of letting their right, by the year, to farmers in neighbouring towns, who in consequence now flock the said common with sheep; and in this such neighbouring farmers must undoubtedly find a very great advantage; as their inclosed lands enable them, first, to breed their own sheep; and, after the pasturage season of the common is over, then to fatten them for the best market. Hence, in proportion as this practice prevails, the common of —— will grow less valuable to the owners, and more valuable to the neighbouring towns.

" That the scarcity of inclosed land renders it impossible for the farmer in —— to breed any kind of cattle with advantage, he being under the necessity of keeping what young cattle he has entirely upon straw in winter; which renders them so poor in kind, that at the time of selling off they are of much less value than those of neighbouring villages.

" That most of the occupiers of farms in —— are obliged yearly to buy hay for their milk cows and working horses, at very high prices, from neighbouring towns, particularly from the towns of —— and ——, and in this respect they will very soon be pressed with still heavier difficulties; the proprietors of the lands of ——, and of the greatest part of ——, having discharged their tenants from selling hay off their respective premises. When it is also observed, that this practice with land owners, of directing the produce of farms to be consumed upon the premises, is become pretty general, the approaching distresses of the ——— farmers are easily conceived.

That the truth of what has been advanced in respect to the farmers of ——— is very strongly evidenced by their poverty, and frequent removals from the town.

"That even the poor themselves, who may be thought to be the least benefited from an inclosure, will reap very important advantages. The condition of the poor of ———, in the winter season, approaches to wretchedness, and that chiefly from the want of one of the most valuable articles among the necessaries of life. This article is milk: of which, as the above statement of facts would induce us to conclude a scarcity; so the poor of ——— confirm the conclusion by sad experience, and can assure us, that this valuable food, which in most places is the cheapest the poor can have, is not to be purchased in ——— at any price whatever. An inclosure would not only be the means of producing plenty of milk; but also of providing for the poor labourers a succession of employment during the winter season.

"That the owners of common right, notwithstanding the present uselessness of these rights, have been, and in all probability may continue to be, exposed to the expence of law suits from the unjust claims of neighbouring towns; as these claims may be naturally supposed to increase, while the common is in its present state.

"That when it is considered, that your resiants of ——— are unanimous in their petition for an inclosure, not so much as a single common-right owner, land owner, or tenant dissenting, the universal good likely to arise therefrom cannot reasonably be doubted.

"That your petitioners therefore beg leave to entertain the hope, that upon your taking their petition, and the substantial reasons on which it is founded, into consideration, you will, in conformity with your known readiness to advance the good of the community in general, be pleased to advance that of your township of ——— in particular, by consenting to the so much wished for inclosure. And your petitioners will ever pray, &c. &c." (Signed by the petitioners.)

Archdeacon Paley, in his political works, very justly remarks, that "there exists, in this country, conditions of tenure, which condemn the land itself to perpetual sterility. Of this kind is the right of *common*, which precludes each proprietor from the improvement, or even the convenient occupation of his estate, without (what can seldom be obtained), the consent of many others. This tenure is also usually embarrassed by the interference of *manorial* claims, under which it often happens, that the surface belongs to one owner, and the soil to another; so that neither owner can stir a clod, without the concurrence of his partner in the property. In many manors, the tenant is restrained from granting leases beyond a short term of years; which renders every plan of solid improvement impracticable. In those cases the owner wants, what the first rule of rational policy requires, "sufficient power over the soil for its perfect cultivation." This power ought to be extended to him by some easy and general law of enfranchisement, partition, and inclosure; which, though compulsory upon the lord, or the rest of the tenants, whilst it has in view the melioration of the soil, and tenders an equitable compensation for every right that it takes away, is neither more arbitrary nor more dangerous to the stability of property, than that which is done in the construction of roads, bridges, embankments, navigable canals, and indeed in almost every public work, in which private owners of land are obliged to accept that price for their property, which an indifferent jury may award."

The person who does not feel the justice of the Archdeacon's remarks, corroborated by the tenor of the foregoing petition, will not be convinced, though we were to give line upon line, and page upon page, in favours of the utility, nay the necessity, of a general division-bill for the whole kingdom being immediately passed.

The next thing we have to state as an obstacle to improvement, is the payment of tithes in kind. We shall here only remark, that the clergy in general are favourable to a commutation, being sensible, that in many instances the payment of this tax in kind, is detrimental to their interest. While the rough hardy collector insists for his full tenth, the quiet good natured clergyman, who studies "if it be possible, to live in peace with all men," is imposed upon in many respects. In short, the payment of tithes is a tax upon industry, for it operates in direct proportion to the merit and abilities of the farmer; and England is almost the only country in Europe, where they are rigorously exacted.

Under the second head, we state want of leases as a great obstacle to improvements. The person who expects land to be improved by a tenant at will, has no knowledge of the human character. Can he be expected to improve, who knows, for certain, that if he were to make improvements, his rent would be increased proportionally. A single fact is worth a dozen of arguments; we therefore give an extract of a letter from a worthy friend in the West Riding, who experiences, to his cost, the truth of what we are now mentioning.

"I delayed writing till I could inform you what was done respecting my farm, and I am now sorry to inform you, that the rent is advanced considerably. It was but laſt week I was informed of the value put upon it by the land doctor; and I have not ſince had an opportunity to ſee my landlord, who I believe keeps out of the way on purpoſe, for fear of being reminded of his former promiſes. I think I mentioned to you, that he once came to my houſe and encouraged me to go on with ſpirited management, aſſuring me that no advantage ſhould be taken thereof. Relying on this promiſe, I manured heavier than ever, and carried on every other improvement in my power; but alas! to my ſorrow, I have been completely deceived. When I had got my farm in tolerable order a perſon was ſent over it, who was a total ſtranger to the management I had practiſed: it was in the autumn, a very growing time, vegetation being rapid after a dry ſummer; my new leys, of which I had a great deal, were covered with a rich verdure, even upon poor ſoil, and the valuation was made according to appearances, without taking into account the extraordinary expences I had laid out in their improvement. I was from home the day my farm was viewed, and was thereby deprived of an opportunity of explaining my management to the inſpector. Indeed, if I had been preſent, I do not know if it would have been of any ſervice, for he charges ſo much per pound for valuing, and the more he advances, the more he receives. In conſequence, you may ſuppoſe my rent is advanced conſiderably. This is fine encouragement to agriculture! I pay intereſt for my own money, and taxed to boot for the improvements I have made. To mend the matter, we are to have no leaſes, and to quit at ſix months warning. I have had a converſation with Mr ———'s man of buſineſs, and frankly told him the injury he was doing, both to the eſtate and the public by ſuch proceedings, but it availed nothing; preſent profit is their object, and they look no farther. This is a pretty lesson to Sir John Sinclair;—he may toil, and scheme, and plan, as long as he lives, at the head of the Board of Agriculture, but it will go all for nothing, as long as gentlemen persevere in these methods of setting land."

The limitations upon management, we also consider as an obstacle to improvements. If the tenant is not allowed to exercise his own judgment, *how* is he to make improvements? If he goes on from one year to another in a beaten course, no alteration can take place in rural œconomy. The same rotation of crops must necessarily be observed, which, in fact, reduces the farmer to the condition of a non-entity, so far as respects the arrangement of his crops. At best, he cannot be considered as superior to the proprietor's steward; nay, he is in a worse situation, for he has the burthen of paying the rent, without being allowed the privilege of exercising his own judgment as to the method of working the ground he possesses. The restrictions or covenants entered into in the West Riding, between landlord and tenant, have been handed down from father to son, for more than a century past; and, if they are suffered to remain, the husbandry of the district will appear in a similar state as at present, to those who survey it a century afterwards.

Another thing prejudicial to the interest of practical husbandmen, and consequently an obstacle to improvements, is the prohibition which the most part of the leases contain against assigning or sub-setting of land. Some writers have lately gone so far as to assert, it is now an understood principle at common law, that unless there shall be a special covenant in the lease to that effect, the farmer can neither assign nor sub-set; and maintain, that the principles upon which this rule has been eſtabliſhed, are grounded in good ſenſe and ſound policy.

It gives us always pain to notice any attempts to place the farmer in a more dependant ſituation than other claſſes of the community, and we cannot in this caſe diſcern the ſmalleſt advantage which proprietors derive from enforcing the prohibition we have mentioned; on the contrary, we are of opinion, that ſecurity for payment of rent, and performance of other obligations, is augmented by ſub-ſets, in a manner ſimilar to what is gained by the holders of bills, who procure a number of indorſers. It is therefore evident, that the advancement of their intereſt is not the reaſon why this liberty is denied to the farmer, and that the cauſes of it muſt be traced to ſome other ſource.

The prohibition againſt ſub-ſetting appears to be a remnant of feudal tyranny, retained by proprietors after the cauſe which introduced it was removed. During the feudal ſyſtem, which put a ſtop to every improvement in Europe, and converted its inhabitants into innumerable hordes of ferocious plunderers, the energy of the huſbandman was totally cramped, he being attached to the ſoil, and equally the property of his maſter, as his cattle and implements of labour; but in proceſs of time, during the progreſſion of power and privilege from the feudal baron to the crown, the cultivators of the ſoil having obtained their liberty, it became neceſſary to ſecure their reſidence on the land, by a ſpecial covenant or clauſe in the leaſe; for, in the feudal ſtate of ſociety, the importance of proprietors did not conſiſt in having a large and well cultivated eſtate, but in the number of armed men they could bring into the field, to face the enemy, or plunder their neighbours; they were therefore particularly intereſted in the perſonal abilities of their tenants, who paid a very conſiderable portion of their rent by perſonal ſervices, in which ſtrength of body, courage, and patience to endure fatigue, were eſſential qualities; and as mankind differ much from one another in theſe qualities, when a proprietor got a vaſſal or tenant on whom he could rely, he took care to keep him on his ground, by prohibiting him from ſub-ſetting, leaſt he ſhould get a weaker or leſs courageous follower in his place. The policy of the country being now happily changed to the better, and feudal ſervices legally aboliſhed, it appears ſurpriſing to us, that this remnant of the ſyſtem ſhould be ſuffered to remain, more eſpecially when it is actually prejudicial to the landholder himſelf.

The ſub-ſet of a farm muſt take place either from the want of capital, knowledge, or induſtry, in the original tenant; therefore, prohibiting ſub-ſets is a certain obſtacle to improvement. If the old tenant is unable to cultivate his land in a perfect manner, it is obviouſly beneficial to the ſtate, that he ſhould be allowed to transfer his right to another, who may be poſſeſſed of capital and knowledge ſufficient for making it produce more abundant crops. On the other hand, the ſub-ſet of a farm may be granted by a tenant of ſuperior knowledge and induſtry, who has laid out great ſums in improving, incloſing, and manuring the ground he poſſeſſes, and who may afterwards wiſh to ſub-ſet it to another of leſs induſtry, knowledge, aud capital, that he may be at liberty to go in queſt of other fields that require extraordinary exertions for improving them. In either caſe, the induſtry of the huſbandman is fettered, and the improvement of the country repreſſed, by the preventing of ſub-ſets, while the intereſt of the proprietor, inſtead of being injured, is rather promoted by the change. In the firſt place, his rent is better ſecured, not only from a more valuable ſtock being generally put on the premiſes, but alſo from the guarantee of the former tenant, who ſtill continues bound for the rent. Secondly, owing to the superior cultivation of the farm, its value is more accurately ascertained by the increased produce; and the proprietor, at the end of the lease, stands a better chance of receiving an adequate rent for it, than if it had remained in the hands of the former occupier. In support of these propositions, we could give many instances of sub-sets, which have produced these beneficial consequences; but we humbly apprehend, that particulars are unnecessary, and that no other reason can be offered for the general aversion which proprietors entertain against sub-setting, but the one we have above mentioned.

We shall now consider this subject in another point of view: Let us suppose a person in possession of a valuable lease, upon the faith of which he procures an extensive credit among his neighbours. In the case of his failure, which, from unforeseen circumstances, may happen without his creditors being aware of it, ought not these creditors to have power to bring his lease to market, and to sell it for their reimbursement? Most certainly they ought. They lent him money, or intrusted him with goods upon the faith of that lease, and reason and equity say, that every part of his property should be attachable by them.

Again; let us suppose the case of a farmer dying, leaving a widow and young family, unable to manage the farm in a proper way and manner, is it not their interest, nine times out of ten, that the farm should be sub-set? Friends are too often indifferent; servants are careless when not looked after; and so it happens, that a lease, which under different circumstances, would have proved a beneficial one, during a minority, turns out to be a bad concern. Few landlords, we believe, in the last case, would refuse their consent to a sub-set; but we contend for it as a right, not as a favour, those who sub-set being at all times liable for implementing the obligations of the leafe, as much as if they were in actual poffeffion.

Few meafures would be more beneficial to agriculture, than placing the cultivators of the ground upon the fame footing with thofe who are employed in trade and manufactures. They have hitherto remained in a much more dependant fituation, although their ufefulnefs to the ftate may, in feveral refpects, be confidered as greatly fuperior. . We therefore recommend, that a fimilar liberty over their own property fhould be granted to them, as is poffeffed by other claffes of the community, which cannot, in any fhape, prove injurious to the land-owner, and would contribute in a material manner to the improvement of the country.

These are the leading obftacles to improvements, and unlefs they are removed, we are confident no material alteration can take place in the hufbandry of the diftrict. The legiflature only can remove the two firft; as for the others, the proprietors, if it were only for felf intereft, ought without lofs of time, to change the nature of the connection betwixt them and their tenantry. If they wifh to draw the utmoft value of their property, it can only be done by giving free and open leafes, without which the tenantry upon every eftate, there, or elfewhere, will be carelefs and indifferent about improvements. Without a leafe, if they make improvements, they are liable to be taxed upon that account, and made to pay intereft in proportion to the money they have expended in improving their farms; and, under limitations of management, they cannot ftep out of the path marked out by their leafes, which may in all probability be an hundred miles further about, than the road travelled by their brethren under different circumftances.

If reftrictions upon management are neceffary, they can only be those of a negative kind. The farmer may, with fume fome propriety, be told what he is *not to do*, but to prefcribe rotations, which feafons or circumftances may render impracticable, or unprofitable to be executed, is detrimental to his intereft. A good farmer, if he can help it, will never have his land in bad order or in an exhaufted ftate, becaufe he knows in thefe cafes he muft hurt himfelf. A bad farmer, tye him up as you pleafe, will always be a bad farmer, and that for the beft reafons in the world. He is ignorant of his bufinefs, and cannot conduct a fingle operation with judgment and wifdom.

" Oh ! but," fay the covenanters, " that's the very reafon we reftrict him; we tell him when he is to make a fummer fallow ; how many crops he is to take ; and in many cafes, even what thefe crops are to be." Do you fo gentlemen? Can you teach him to make his fallow clean? Or, if you are able to teach him, can you force him to put it in that condition ? And, if he follows your rotation, and fows the very crops marked out, are you certain the different operations of plowing, fowing, harrowing, and reaping, will be executed with propriety ? No, this is impoffible : A bad farmer will conftantly act in character, reftrict him as you will ; while he who knows his bufinefs, give him the moft unlimited powers, will always labour in fuch a way as not to injure the ground, becaufe he knows he cannot do this without injuring his own intereft.

We might have mentioned the fmallnefs of the Weft Riding farms as an obftacle to improvement, were we not fatisfied, that, in a manufacturing diftrict, fmall farms muft neceffarily prevail. In thofe parts which are at a diftance from the manufacturing towns, where farming is a bufinefs, they are of a greater fize, and much better managed. Not that we think good management cannot be practifed upon fmall farms ; quite the contrary, as the prefent ftate of Flanders bears teftimony. Where a country is already improven, small possessions may be very proper, provided the occupier works himself, as they are not sufficient to keep a man idle, and he has not others to oversee ; but in every country where great and substantial improvements are to be introduced and carryed on, unless they are executed at the expence of the proprietor, there is a necessity of having farms of a large size, so as men of capital and knowledge may be stimulated to enter into the profession. That this is a fact, the present state of husbandry in the different parts of Britain sufficiently demonstrates.

CHAPTER XVIII.

MEANS OF IMPROVEMENT, AND THE MEASURES CALCULATED FOR THAT PURPOSE.

AGRICULTURE is the parent of all the arts, and the practice of it may be confidered as a ftandard for the flourifhing of others. It has for fome years paft been a principal object in the feveral governments of Europe, to frame laws and regulations for its encouragement; and the eftablifhment of a Board for promoting Agriculture and Internal Improvement, fhews it is not neglected in our own country. We have, in the foregoing parts of this work, fubmitted to the confideration of that Honourable Board, a ftate of the hufbandry in this Riding, and alfo pointed out, for their information, the principal obftacles which are in the way of further improvements: we now proceed to fuggeft how thefe obftacles may be removed, and what alterations ought to be introduced into the hufbandry of the diftrict.

The improvements we fuggeft are:

1ft, That the nature of the connection betwixt the landlord and the tenant fhould be changed, and that leafes of a proper duration fhould be granted.

2dly, That the arbitrary and injudicious covenants generally impofed upon the tenantry, fhould be difcontinued, and conditions more favourable to improvements fubftituted in their ftead.

3dly, That tithes should be commuted.

4thly, That a general bill should be passed by the legislature, for the division of the common fields and waste grounds.

These are the leading means of improvement; without which no material encouragement can be given to the husbandry of the district. In hopes that the Board of Agriculture will consider them in the same light, we proceed to recommend,

5thly, More improved rotations of crops.

6thly, Breaking up the old pasture fields, and frequent changes of corn and grass.

7thly, Drilling and horse-hoeing beans and turnips.

8thly, Planting the waste lands which are improper for cultivation (*a*).

Many other articles of lesser importance might be added, but as most of them are already noticed in the foregoing parts of this work, we shall not now enter upon them.

1st. That all lands should be let upon leases.—We have often had occasion in the preceding pages to shew the baneful consequences attending the want of leases, and how few real improvements will ever be introduced into the practice of agriculture, so long as the farmer has no security for enjoying his possession more than one year. We therefore recommend, as a necessary step, to encourage good farming, that leases should be granted of a proper duration (*b*). This would not only operate in favour of the farmer, but would likewise be the means of increasing the rent-roll of the proprietor; for no man will ever pay so much for an acre of land, while he is removable at pleasure, as when a kind of permanency is granted him. Upon all lands already in a state of cultivation, we think 19 or 21 years are very proper terms for the continuance of a lease. They afford the farmer time and opportunity to make improvements, and to receive a proper return for the money ſo laid out, without depriving the landlord, farther than neceſſary, of any advantages that might ariſe to him from a progreſſive increaſe in the value of his grounds. Without this ſecurity, no farmer will engage in any expenſive or ſpirited management; and the diſtrict will remain unimproved to its utmoſt extent.

2dly, If leaſes are granted of a proper duration, the neceſſity of impoſing reſtrictive covenants upon the tenantry will, in a great meaſure, be removed. If it is thought neceſſary for protecting the landlord's property to reſtrict the tenant for the three laſt years of the leaſe, we ſee little harm that would ariſe either to individuals or the public from that meaſure (*c*). Our ideas of a leaſe are, that it is a mere bargain betwixt landlord and tenant, wherein the former, for a valuable conſideration, to be paid annually, conveys over to the latter, all his right in the premiſes for a ſpecified number of years, and that during their currency, it ought to be left to the wiſdom and abilities of the farmer, to manage the land in ſuch a manner as he may think moſt proper for enabling him to make good his engagements to the landlord. If leaſes were granted upon theſe principles, a great deal of unneceſſary trouble would be ſaved to both parties, improvements would increaſe with rapidity, and the peace, comfort, and happineſs of that uſeful body of men, the farmers, would be materially promoted.

Our opinions upon the clauſes that ſhould be inſerted in a leaſe are plain and ſimple, and we beg leave to ſtate what the heads of theſe ſhould be:

The landlord agrees for a ſpecified rent, payable at the terms of Candlemas, Whitſunday, and Lammas, after the crop is off the ground, to ſet ſuch a portion of land for 21 years; and to put all the houses, offices, and inclosures upon the premises in habitable and fencible condition.

The tenant agrees to pay rent as aforesaid, and to forfeit his lease if payments are not made within six months after they fall due, with interest for the intervening time; to manage the land in a husband-like manner, and not to sell straw (*d*) or dung off the premises; to support all the houses and fences during the continuance of the lease, and to leave them at its expiration in a habitable and fencible condition (*e*); to leave *one fourth* of the farm in grass at least three years old, and likewise a *sixth part* of the remainder as fallow to the in-coming tenant, upon allowance being made him by valuation of neutral persons; and if any dispute arise betwixt the parties during the lease, or about the situation of the houses and fences at the conclusion, that the same shall be referred to the determination of arbitrators mutually chosen (*f*). If land was let agreeably to this method, the management of an estate would comparatively be an easy task to what it is at present; and while no injury was done to the landlord, the condition of the farmer, from being uncertain, would be respectable and happy (*g*).

3dly, That tithes should be commuted.—After the restriction imposed upon us by the Board respecting this article, it would be improper to add more, than that the real interest of the country is concerned in having them regulated in one way or other as soon as possible.

4thly, Division of the common fields and waste grounds.—After what we have already mentioned, Chapter XII. it is almost unnecessary to say any thing further on this subject; but it is of such importance to the public, that we will readily be excused for stating a few additional arguments in support of this measure.

The proper way of discussing a question of this nature is to enquire, whether the holding of land in commonty, or severalty, is most conducive to the public good? or, in other words, whether the ground is most productive under the one tenure or the other? It is the improvement of the country which the Board ought to have in view, and not the augmentation of this or that man's property; and, even supposing that private rights may be partially injured, yet if a general division of these common fields and wastes will increase the quantity of corn or live stock, the interest of the country is thereby promoted. Now, as no land can be improved when lying in commonty, it follows, that putting it in that state which allows the proprietor to cultivate and manure it as he pleases, must be a necessary measure, and that the object justly deserves the most serious attention from a Board, expressly established for encouraging internal improvement.

The common fields cannot be considered as yielding one half of their natural value, in the way they are managed. They are exhausted by long and continued tillage; the same rotation of crops has been followed out, for time immemorial, and, in their present situation, improvement is impracticable. To remove every obstacle to their melioration, is the duty of the legislature; and experience has ascertained, that without one general bill, which must operate upon all, and which in many instances will cut the knot that cannot be loosed, the public interest must continue to suffer from the unproductive state of these lands.

The situation of the waste lands reflects shame on the policy of England, for, while they continue in their present state, the country derives scarce any benefit from them. Many of them are susceptible of great improvement, providing the owners were emancipated from those legal obstructions which have hitherto prevented them from cultivating what ought to be their own property. If the waste lands of Britain were cultivated in a wise and judicious manner, they would be of more solid value to the nation, than the whole of our West India possessions; and it presents a melancholy picture, that while we have eagerly contended for the possession of distant countries, we have carelessly neglected the melioration of at least one fixth part of our home territories, which were undoubtedly of much more importance.

5thly, Introducing more approved rotations of crops.— If leases of a proper duration are not to be granted, and if the practice of binding up tenants with restrictive covenants is continued, it would be perfectly unnecessary to suggest any improvement in the mode of cropping the ground, as however willing the farmer may be to adopt new practices, he is in a manner prohibited from doing so by the conditions under which he holds his possession. But entertaining sanguine expectations that these obstacles to good husbandry will soon be removed, we proceed to point out such alterations as, in our humble opinion, are proper to be introduced into the husbandry of the district.

Viewing the present state of farming in a general manner, it appears that the land in the West Riding is cultivated in two separate and distinct ways, and not managed so as to make improvements in one branch contribute to the advantage of the other. The fields which are laid down in grass, continue in a state of pasture for a greater number of years than is necessary for refreshing them, after being exhausted with corn crops; while the fields kept under the plough are hackneyed and worn out by successive crops of corn, without receiving any collateral assistance but what is given them by fallow and manure, with some passing clover crops (*b*) (*i*).

We consider it as essential to good husbandry, to connect these different systems, and that the ground in no other way can be kept in a perpetual state of fertility, or made to produce its utmost value. While we decidedly condemn the keeping land exclusively in grass, we as warmly reprehend the contrary extreme of persisting uniformly in raising crops by the plough. The last named practice may be said to have necessity upon its side, whereas no excuse can be offered as a palliation for the other.

Upon the supposition that the system of keeping lands continually in grass will be departed from, and that the farmer will be allowed to cultivate his fields in such a way as he thinks most beneficial, we shall give our opinion upon the most advantageous method of cropping a farm; or, in other words, shew how it may be kept in continual good order, so as to enable the possessor to pay the highest rent, while at the same time it is understood he is to receive a proper recompence for the expence and labour he is at in cultivating it.

The first thing that is absolutely necessary in farming land well, is to lay it clean and dry. Where land is foul, carrying either quickens, or other weeds, it is impossible artificial plants, such a corn and grass, can thrive. The ground is bound up, and the food, that should go for the support of the plants sown by the husbandman, is exhausted by these natural inhabitants of the soil. The crops, therefore, are scanty, being stinted in their growth, and inferior in their quality. Every good farmer will therefore use his earliest efforts to make his land clean. This he will do by complete summer fallows, or by fallow crops adapted to the different soils he possesses; and having once accomplished his purpose, he will studiously endeavour to preserve it in the same husband-like order.

That land may be kept clean, a powerful assistant is gained from having it previously laid dry, or in a proper situation for carrying off the superfluous water that falls upon it from the clouds, or arises from the veins of the earth by springs, or from being situated upon a wet spongy bottom. This is done by ridging the land sufficiently high, for defending it against falls of rain; by casting out the water furrows, provincially called "*griping the land*;" and by digging hollow drains, which, when covered, carry off the superabundant moisture, and occasion no loss of ground. These two things, laying the land dry, and keeping it clean, are in the power of every farmer, although they are more difficult to execute in some situations than in others; but there is another principle requisite for bringing farming to its greatest improvement, which is to keep the land also rich; this is often not in the power of the best farmer to command, and must in a great measure be regulated by local situation, or by the particular quality of the land he possesses.

Having premised these things, which we consider to be the fundamental principles of good farming, we shall now state our ideas upon the way in which a farm should be managed, and the particular crops most advantageous to be raised upon different soils.

Upon all gravelly, sandy, and sharp soils, allowing there may be a degree of hardness in them, we recommend the turnip husbandry to be assiduously practised. Upon such soils turnips may be introduced every fourth or fifth year. In those parts where cutting the clover crop for hay is attended with profit, they come in with propriety every fourth year; but in many situations, we judge it more advantageous, in place of sowing the barley crop with red clover, to sow it with white clover, trefoil, and rye grass, and to pasture it for two years with sheep (*k*); as red clover is found from experience not to answer well, when too often repeated. This gives the ground a proper cessation from tillage, invigorates its powers, prepares it for carrying a weighty crop of oats, with very little collateral aſſiſtance from manure, and allures nature with variety, which is always agreeable.

A farm managed in this ſtyle, will conſiſt of five breaks or parts. *1ſt*, Turnips. The firſt half of the turnips that are conſumed, to be ſown with wheat, the laſt half with barley, and both ſown with graſs ſeeds; paſtured the third and fourth years with ſheep, and limed if thought neceſſary upon the ſward, or with turnip crop, as is thought moſt advantageous; fifth year, broke up for oats, which will always be found in this way a profitable crop.

Land of the above quality, managed in this manner, (and the Weſt Riding land, from being moſtly incloſed is admirably calculated for it), will pay both proprietor and farmer better than moſt other ſoils. Expences of management, which is a great conſideration, are comparatively trifling; and no foreign manure, when once the rotation is properly arranged, will ever be needed.

The ſame mode of cropping, although not with equal advantages, may be carried on upon all loamy ſoils, unleſs they have too great a portion of clay in them; but if the farm is of a mixt nature, and has both dry gravel and loam in it, we recommend that the turnip break may be ſo arranged as to take in both ſoils, and that thoſe upon the loam be eaten off firſt, and the land ridged up immediately, which will both lay it dry, and afford opportunity for correcting the ſtiffneſs and adheſion it may have contracted, by the preſſure and poaching of the ſheep.

Upon land where clay is a principal component part, or where the bottom is wet, we cannot recommend the cultivation of turnips at all, as often the profit gained from them is loſt upon the following crops. The ſame objection holds againſt cabbages, rape, or any other plants that are to be eaten off in the winter months (*l*).

Lands of this nature are more difficult to manage, than those already described, and from being cultivated at a greater expence, are never able to afford so much rent to the proprietor, allowing the crops raised upon them should be as productive as those raised upon the dry soils. Beans is the only crop that can be introduced for cleaning the ground; but although these are an excellent assistant, they can never preclude a complete summer fallow from being absolutely indispensable.

A farm of this sort ought to be divided into seven breaks or parts, and the following rotation is in our opinion most adviseable:

1. Fallow, with dung (*m*).
2. Wheat.
3. Beans, drilled and horse hoed.
4. Barley, sown with grass seeds.
5. Pasture.
6. Pasture.
7. Oats (*n*) (*o*).

In order that a proper season for sowing the wheat upon such soils may not be missed, we recommend it to be sown by the middle of September. Crops early sown, though they never yield proportionably to their bulk, yet are generally most productive per acre; and it is an important matter upon all clay soils that the seed should be put in dry. Wet harrowing not only dibbles in the seed beyond the power of vegetation, but also poaches and binds the land, by which the plants are prevented from stooling, or tillering, and gives encouragement to the growth of any quicken that may be left in the ground. The beans should get two furrows, the first across, and particular pains should be afterwards taken to water-furrow the land. The seed should be put in as early as possible, after the land is in a situation to stand a second ploughing, as the quantity and quality of the crop depend much upon an early feed time. Barley may be sown after two furrows; for if proper attention has been given to the bean crop the preceding year, the ground will be in good order, and spring-ploughing upon clay land is always critical.

In the above rotation, a proper arrangement of labour is made for the whole season. The part destined for wheat is prepared during the summer months; the first furrow is given for the beans as soon as the wheat is removed; next the barley land is fallowed down; then one of the pasture fields ploughed for oats, and the first furrow given to the next year's summer fallow, which concludes the winter operations. In the spring, begin with the bean feed, next sow the oats, and finish with the barley feed; which finishes the work of the season, and allots to each particular period a proper quantity of work, without hurrying too much at once; which ought always to be regarded, especially upon clay soils, as a material object.

The thin poor clays are the most difficult to farm of any kind of land, and nothing can be done upon them to purpose, without the aid of a greater quantity of manure than what can be raised upon the premises. At the same time it is perfectly unnecessary to lay a great quantity of manure of any kind upon them at once, for they possess a quality so corroding, that the aid, thereby given to vegetation, is soon wasted and lost. Where local situation will allow, we recommend such land to be kept in four breaks, and cropped as follows:

1. Fallow.	3. Pasture.
2. Wheat.	4. Oats.

This rotation will pay very well, if manure can be got sufficient to cover the fallow break. The pasture should lie only one year, as land of this kind does not improve in grass; and the oats will be found better, in such a case, than if the grass had been older.

In order that the rotations here recommended may be followed out to the greatest advantage, it is absolutely necessary that particular attention should be paid to the summer fallow, or to the turnip crop substituted in its stead. If any error is fallen into in this stage of the business, the after crops are consequently injured. Broad cast turnips can never be considered as a fallow crop, no hand-hoeing being equal to cleaning with the plough.

6thly, Breaking up the old pastures, and frequent changes of corn and grass.—This subject we have already discussed, p. 114, and here will only add, that it is ascertained by facts, that a leguminous and culmiferous crop alternately, affords the greatest possible return from the soil. The old pastures of Yorkshire would be greatly benefited by being broken up, as they are in too many places stocked with rushes and other trumpery, while the surfac in others is over-run with moles, and consequently in a state disgusting to the eye, and prejudicial to the growth of good grasses.

The celebrated Walter Blyth, in his "Improver Improved," printed in the year 1652, and now a very scarce book, seems to have considered this subject in a similar manner. He says, (page 95. of that work,) "There is another extreme which men wedded to their selfe profit, hugg in their very bosome, which is so much to their hearts content, that they never look what may make most profit to the publique, or good of the commonwealth, themselves, or posterity. He is seated in way of feeding and grazing, with a constant stock of breeding, and let his land be fit for one, or fit for another use, he matters it not; he hath received a prejudice against plowing, partly because of the toyle and charge thereof, and partly because some men have plowed their land so long as they have impoveriſhed it much, and ſome men ſo long, as it is poſſible it may be many years before it ſoad completely; and therefore, let it be dry or moyſt, ſound or rotten, ruſhy or moſſey, fenney or run over with a flag graſſe, or ant hills, moſſure or wild time, let it keepe more or leſſe, hei'l not after; tell him, ſir, it will yield abundance of gallant corn to ſupply the whole country, and raiſe great ſums of money to your purſe; and afterwards, (if you plow moderately), it may keepe as many cattle, nay more, yet nothing takes with him, he will have no encloſure plowed by no means."

Again he ſays, page 101, "I once held a piece of land worth nine ſhillings an acre, and no more to graſe: I gave fifteen ſhillings to plough; it was great lands, as great balks betwixt them, full of your ſoft ruſhes, and as high ſome of them as an ordinary beaſt, and lay very wet. The land, conceived by me, not able to beare barley, nor never would, it was ſo weak, barren, and cold; and the neighbours, very able huſbandmen round about, ſo diſcouraged me, (out of their love unto me), as that they deſired me to forbear tillage of it, becauſe it would never anſwer ordinary coaſt beſtowed upon it, nor be worth an old grazing rent to plough; and that they cleared to me, by very clear evidence, as they conceived, affirming, that the land next unto it, but a hedge betwixt, which was farre better land, (and indeed ſo it was, very neare as rich again), huſbanded by very able huſbands, the beſt in that country; and that land, good barley-land, yet never anſwered the pains and coſt beſtowed, yet I reſolving to make a full triall thereof, I ſet upon it ploughing, harrowing, ſpading, and dreſſing, (for indeed I made harrows on purpoſe alſo) of divers ſizes, it coſt me about fifteen ſhillings an acre the two firſt crops, the very dreſſing of it; and for theſe crops, being but of oates, I could have had five pound an acre, being offered it by an oate meal-man of himfelf, though never afked, growing upon the ground; nay fix pounds an acre, if I would have fold it, which is a vaft rate for oates in the middle of the nation. And indeed I found the ground fo poore, that it would not beare barley, for I tryed fome acres of the beft land in it, but it was not worth an acre of my oates; and after ploughing, I gave the old naturall rent, as it was ever fet at, or really worth, and that for many years; and the land is better, lyeth founder, warmer, and both yields more milke, fummars as many cattle, and winters farre more, and feeds better than it did before, without any other coft beftowed, and the very firft yeare I layd it down after ploughing, it kept me more cattle, and better than ever it did before, and will continue better for it for ever after."

7thly, Drilling and horfe-hoeing beans and turnips, is an improvement which we earneftly recommend; but while we are eager for drilling thefe two crops, we cannot recommend this practice to be ufed for raifing other grains. Wheat, barley, and oats are found both better in quality and quantity when fown broad-caft (w); and the reafons are thefe—When drilled, they are much expofed to the weather, and are liable to be broke down and injured by every gale. Befides, they tiller or ftool as long as any interval is left, which neceffarily caufes the grain to be unequal (17).

When the land is judicioufly prepared, and due attention given to the cleaning of beans and turnips, the neceffity of a fummer fallow is in a great meafure fuperfeded. Many foils, undoubtedly, cannot be kept in order, unlefs they receive a complete fummer fallow; but it is as certain, that if due care is ufed in the working of thefe crops, a frequent repetition of this practice will be unneceffary. Wherever the ground is in order to produce a good crop, it ought not to remain unproductive for a ſeaſon; but, unleſs drilled crops are frequently reſorted to, and judiciouſly cultivated, ſummer fallows muſt intervene oftener than is conſiſtent with the intereſt of the farmer. No part of what we have ſaid in favours of the drilling and horſe-hoeing huſbandry, is meant againſt the practice of ſummer fallowing, when the condition of the ground requires it. Upon every variety of clay ſoils, good management cannot be carried on without it. We only contend, that drilling and horſe-hoeing certain crops, will enable the farmer to extend his rotation much farther, than if he were conſtantly to ſow in the broad-caſt way.

We have reaſon to ſuſpect that the intricate nature, and expenſive coſt of drill machines, have deterred a number of farmers from adopting this mode of huſbandry. We venture to affirm, that the ſimpleſt machines are the beſt, and that a bean drill, which may be made by every common wright for 12s, and a turnip one for about double the price, ſowing one row or drill at a time, will be found of more real utility, than all the expenſive complicated patent machines in the kingdom.

Planting the Waſtes.—If the waſtes were divided, we are fully convinced that much improvement might be made by planting Scots firs and larches upon many parts of them. Theſe kinds of wood are at preſent held in little repute, and are indeed ſcarcely known in the Weſt Riding. As a great deal of fir wood is at preſent imported from the Baltic, they might in time render that, in a great meaſure, unneceſſary. They would anſwer for roofing cottages, for fences, and many other uſeful purpoſes. The ſubject deſerves attention, and we are humbly of opinion, that the far greater part of the moors, in this diſtrict, can never be improved in any other way.

Scots firs and larches are the hardieſt of all foreſt

trees, and will thrive upon the moſt barren ſoils. They ought to be planted pretty thick, ſo as to afford ſhelter to each other, and great care ſhould be uſed to thin them as often as neceſſary. The very prunings of them would, in a ſhort time, be equal to the preſent value of ſeveral of the waſtes, and when the trees were arrived at that ſize, as to be ſit for fencing, &c. the yearly return would conſequently augment, while, at the ſame time, the graſs within the plantations or woods, would be of greater value than when it remained in its original ſtate. This has actually happened in our own country, where plantations have been made upon ſuch barren ſoils as we are here mentioning.

It ought to be a material object with every well regulated government, that no part of its territory be allowed to remain unproductive, but that every acre of it ſhould be employed to ſome uſeful purpoſe or other. If the ſoil is the capital ſtock of the country, as ſeveral political writers have maintained, the intereſt of the country is neglected when any part of it is allowed to lye in a ſtate of ſterility. There is no part of the earth but what may be adapted to ſome uſeful purpoſe or other; and, as there is a conſtant demand for wood in the Weſt Riding, the proprietors are called upon, both by public and private motives, to plant every acre not capable of being improved by the ordinary methods of cultivation.

Several other things might have been noticed, as contributing to improvement, did we not wiſh to confine ourſelves to the great and leading features. We ſhall juſt hint at a few of them:—1ſt, It would be no injury to the proprietor, and ſave much trouble to the tenant, if all public taxes were paid by the former; beſides, the tenant is very apt to conceive an idea, that theſe burthens are not a part of the rent, but that he is paying heavy taxes, while his landlord is free. We confess, that we would not be fond of signing a lease, which oblig ed us to pay all parliamentary taxes *already imposed, or to be imposed*, which, in the present state of our national finances, might prove a serious business. So, if taxes are a part of the rent, the lessee, under the clause we are alluding to, undertakes to pay an unknown sum for the farm he is to possess, which may, for ought he knows, be much more than its actual value.—2dly, It would be of material advantage to agriculture, that some alteration was made upon the game laws, and that the privilege of hunting was used in a more lenient way. It really shocks the feelings of a farmer, to notice the injuries committed by a parcel of people mounted on horseback, and galloping like madmen after a poor fox, or an innocent hare. We are convinced, that no *real* gentleman will injure the property of the farmer, when engaged in this *diversion*; but, so it happens, that heavy losses are often sustained, by those over whose fields the object of sport happens to run: fences are tumbled down, the sown wheat fields rode over, the young grasses not spared, and, in short, every thing must make way for these sons of Nimrod. It is a poor consolation to the farmer, that he is entitled to damages for these wanton devastations. These can in few cases be estimated, and are as seldom paid. The law, in other cases, does not allow the person who injures his neighbour, to get so easily off. If a house is broke into, or a purse stolen, it is not a compensation in kind that will absolve the culprit from the consequences of his delinquency.

It has been suggested to us, that it would be of public advantage, for the Board to take into their own hands, experimental farms in different parts of the country; and, that if this measure was adpoted upon every variety of ſoil, and the management, for which they are naturally diſpoſed, ſteadily adhered to, real knowledge in huſbandry would increaſe in courſe, and ſubſtantial improvements be rapidly diſſeminated.

Viewing the buſineſs in this light, we think the ſuggeſtion merits the ſerious conſideration of the Board. Example, in moſt caſes, goes before precept, and the moſt obſtinate old-faſhioned agriculturiſt, would be ſharp enough ſighted to his own intereſt, ſo as to change his practice, the moment he perceived a more advantageous one placed before his eyes.

Such eſtabliſhments might likewiſe ſerve other ſalutary purpoſes: They might be conſidered as academies for training up young men to the practice of agriculture, a branch of education too much neglected and undervalued. Practical huſbandry might there be taught in all its branches, from the ploughing of the ground, to the dreſſing of the corn for the market; and, inſtead of the teacher entertaining his pupils with florid theoretical harangues about the paſture of plants, and things of the like nature, he might take them to the field, and, with the inſtrument in his hand, lecture upon the different proceſſes of farm labour. More real advantage would, in this caſe, be derived from half an hour's teaching, than from a whole ſeſſion's attendance upon a college profeſſor.

4thly, It would be very conducive to agricultural improvement, that encouragement was given for increaſing the number of farm ſervants and labourers. This can only be done by amending the poors laws, and by building cottages contiguous to every home-ſtead. At this time, the farmer is apprehenſive of having married ſervants about him, becauſe he knows that a riſe of the poors rates is the certain conſequence. He therefore hires young men, boards them in his houfe at a great expence, and keeps the evil from his own doors as long as poffible. But this does not ferve the public intereft, as marriage amongft the lower claffes is difcountenanced, and the number of operative hufbandmen is decreafed; wages are augmented, and numerous evils occafioned to the community, all which might be leffened, was fuitable attention paid to the objects we have affigned as the caufes.

We fhall now bring this Survey to a conclufion: In the courfe of it, we have endeavoured to defcribe the ftate of hufbandry in the Weft Riding, as is it actually carried on; and this we have done, not only from a minute examination of its different branches, and from the information collected during the time we remained in the Riding, but likewife from the very liberal communications of feveral intelligent gentlemen and farmers, fince tranfmitted to us. It certainly has given us much pain, to be under the neceffity of cenfuring feveral prevailing practices, and to fay fo much againft the nature of the connection which at prefent generally fubfifts betwixt the proprietors and their tenants. Upon thefe matters, we have ftated our fentiments with freedom, becaufe we are fenfible of their importance; and have uniformly acted upon this maxim, that "thofe who are afraid of the public, are not the men by whom the public is to be ferved (r)."

With regard to the intereft of that ufeful body of men the farmers, we have endeavoured to fhew how much their fituation would be meliorated, and the practice of agriculture improved, by the proprietors granting leafes of a proper duration, free of thefe ufelefs reftrictions and covenants that now fubfift in agreements for land, whether annual, or for a greater number of years.

These things we humbly fubmit to the confideration of the Board of Agriculture, and we entertain the fanguine hopes, if the improvements we have fuggefted are fanctioned by their approbation, that this sanction will have great influence in correcting the abuses we have described, and contribute to improve the husbandry of the West Riding of Yorkshire: by which means the interest of the landed proprietor will be augmented, the peace and happiness of the farmer increased, and consequently the public good materially promoted.

NOTES on Chap. 18.

(*a*) I approve of these suggestions. *S. Birks, Esq.*

(*b*) The interest of the landlord seems to require him to grant leases of a moderate length, to tenants of arable farms, in order to encourage them to lay down land and work fallows in the most perfect manner; and often to purchase manures at a greater expence than they can expect to be reimbursed by the ensuing crop. Graziers sometimes expend considerable sums in the purchase of lime, which they lay upon their grass land. The lengths of the several leases should bear a proportion to the respectively proposed expence, and each ought to be settled by the mutual agreement of the parties, adapted to the particular case, and not by any general rule. The length of leases here proposed, seem to be greater than are necessary, in ordinary cases, to enable the lessees to reimburse to themselves such expences as they are likely, or indeed reasonably wished, to incur. Such leases, especially without restrictions, estrange landlords from their estates, who would consider themselves as annuitants, and would set little value upon reversions, which it would be very uncertain whether they should live to enjoy. They would become utterly disinclined to undertake the great works of building, inclosing, and draining, which are much more effectually performed by landlords than tenants; however, where landlords live at great distances from their estates, or are disinclined to undertake these works themselves, it is sometimes adviseable that they should let leases of considerable length to substantial people, upon conditions, and with very strict covenants, that they will execute such works; otherwise these great works would not be performed by either party.

T. York, Esq.

Answer.—It is unnecessary here to enter upon a refutation of what is contained in the above note, as we have already stated our sentiments fully upon the subject, pages 46, 47, 48, and 49. We only add, that the length of the lease we have recommended, is moderate in the extreme; for, upon an arable farm, it just admits the rotation to go thrice round: We are at a loss to conceive what pleaſure any landlord can receive from an interference with his tenants buſineſs, or in what manner he ſhould be eſtranged from his eſtate, becauſe thoſe below him are placed in a comfortable ſituation. Such inſinuations are libels upon the landed intereſt of England, who, it is hoped, will never generally countenance a ſyſtem ſo injurious to the improvement of the country, and ſo detrimental to their own intereſt. *R. B.*

(*c*) It appears right in the nature of things, that every farm ſhould be let at leaſt three years before the expiration of the leaſe of 21 years. In that caſe, the farmer goes on with his ſyſtem without interruption; but if otherwiſe, he has it in his power to injure the farm in a degree, eſpecially if he has any ſuſpicion of being removed. *Mr Culley.*

(*d*) In many places *ſtraw* is one of the moſt valuable commodities the farmer has to diſpoſe of; to debar him therefore, from ſelling it, would be the greateſt abſurdity. The form of a leaſe here given, is liable to numerous objections. *Anonymous.*

Anſwer.—Over the greateſt part of the iſland, the farmer has no materials for dung, but the ſtraw of his crop. To ſuffer him to ſell it, would be abſurdity in the extreme, as the ground in a few years would be utterly impoveriſhed. It would have been obliging if the remarker had ſtated his objections to the form of a leaſe we recommended, in more particular terms. *R. B.*

(*e*) The intereſt of the landlord requires, that he ſhould ſupport the buildings, otherwiſe he would probably find them much out of repair at the expiration of a leaſe, notwithſtanding covenants. *T. York, Eſq.*

Anſwer.—If the tenant is bound by his leaſe to ſupport the houſes during its continuance, and to leave them at the concluſion, in a habitable condition, or in other words, as good as he got them, it does not appear they will fall into the landlord's hands in ſuch a ruinous ſtate as Mr York dreads. But if they are out of repair, the old tenant ought either to put them in proper condition, or make payment of a ſum of money to the landlord or his ſucceſſor, equal to the amount of the repairs which are neceſſary. *R. B.*

(*f*) This definition of a leafe feems to be very erroneous. It rather feems to be a conveyance, &c. with fuch refervations, limitations, and conditions, as are mutually agreed upon by the parties; however, they muft be poffible, and not repugnant to the rules of law. *T. York, Efq.*

Anfwer.—Call it a conveyance; call it what you pleafe; fo long as the tenant implements the articles of the bargain, the landlord can in equity have no right to interfere. *R. B.*

(*g*) Thefe are very proper conditions for a leafe; as, while the landlord's property is protected, full liberty is allowed to the farmer to exercife his abilities and induftry. *A Farmer.*

(*h*) It is fully admitted, that very old pafture would be confiderably improved, by being broke up and brought into a regular courfe of tillage; that the profit arifing from it would enable the occupiers to lay down their old arable lands in a moft hufbandlike manner; and that fuch alternate management in future, would be very beneficial to landlords and tenants; however, it feems fcarce neceffary to add, that this ought to be regulated by very particular covenants, adapted to each cafe. If an uncontrolled power were refigned to tenants, mifchief vaftly exceeding the propofed benefits, would be the neceffary confequence. An apprehenfion of thefe, frequently renders abfentees, and fome refident land-owners, who have not turned their thoughts particularly to the ftudy of agriculture, exceedingly averfe to propofals for breaking up frefh land. Prejudices, and it is admitted that fome take place, fhould be removed by conviction; it is proper that the queftion fhould be thoroughly agitated and difcuffed, by which it will be more generally underftood, and then each individual proprietor, by promoting his own intereft, will promote that of the public; which is, that the lands of England fhould conftantly be in a progreffive ftate of improvement. The tenant hath no permanent intereft in the land; he, of courfe, endeavours to get as much as he can during the term; he often thinks his intereft oppofite to that of his landlord, and exercifes his fkill to bring down the value of the land, towards the latter end of the term, with a view of re-taking upon eafy terms.

T. York, Efq.

Anfwer.—If leafes were granted for twenty-one years as recommended, it would be found, that the intereſt of the landlord and the tenant is the ſame, for at leaſt ſixteen years of the term; and facts warrant us in affirming, that a few ſcourging crops at the concluſion, does not leſſen the *real* value of the property; at the ſame time, we muſt remark, that protecting clauſes were recommended for theſe years. *R. B.*

(*i*) This is a great error in the management of this country, as the old farmers and the ſmall farmers generally travel in this John-Trot road. The old ones will not be convinced, and the ſmall ones cannot practiſe a proper change for want of room.

A Farmer.

(*k*) New lands are ſaid by many to inſure the rott.

Meſſrs S. P. and M.

Anſwer—So will old graſs if the bottom is wet. *R. B.*

(*l*) We have a great proportion of this ſort of land, and experience proves, that turnips upon it are extremely hurtful to the ſucceeding crops. *T. H.*

(*m*) Would it not be better to lay no dung upon the fallow, but reſerve it for the bean crop? *Mr Culley.*

Anſwer.—The uncertainty of getting the dung laid on to the beans, is not to be riſqued. *G. R.*

(*n*) It is believed, that land would ſoon be exhauſted, unleſs it were exceeding rich, or very highly manured, by this courſe. The land would be favoured by leaving out the barley, introducing horſe-hoed beans, and paſture alternately the third year.

T. York, Eſq.

Anſwer.—The moſt part of Engliſh landlords labour under a kind of nervous affection, in caſe their land ſhould be deteriorated. The above courſe of cropping, is one of the moſt approved, for ſoils of the quality alluded to. How Mr York is to mend the matter by leaving out the barley crop, I cannot conceive.

R. B.

(*o*) The greater part of clay lands in this country, is too poor to bear this rotation of crops. The barley would be worth nothing, unleſs freſh manured, nor would the ſeeds come to any perfection. I would rather prefer sowing the seeds with the wheat, then pasture one or two years, and afterwards sow wheat and oats, or oats and wheat. *T. H.*

(*p*) Repeated experiments on my farm, have proved the reverse of what is here stated. *W. D.*

(*q*) I have practised what we call scurbaking several years, and find it to answer very well. The ridges are made up at about the same distance as for planting potatoes, and the seed sown broad-cast. I then horse-hoe and hand-hoe, and have had by far better crops in the open fields, than any of my neighbours, by broad-cast without ridges, and my land much cleaner, and fitter for succeeding crops. However, I approve of the distance here recommended, of twenty-four or twenty-seven inches, and of plowing betwixt with the small plough; but I think it may be done in this way, sown by the hand in broad-cast, after the ridges are formed, as well as by a drill. *A Yorkshire Farmer.*

Answer.—By no means so regular as the drill. *C. R.*

(*r*) However unpleasant many of the observations in this report may be, to those concerned in the abuses which are pointed out, I beg leave to recommend the work to the attentive perusal of the Members of the Board of Agriculture. *W. D.*

A gentleman who signs himself " A Yorkshire Farmer," and to whom we have been under great obligations, introduced his remarks in the following words:

The annexed marginal remarks, are humbly submitted to the consideration of the Board of Agriculture, and, should they prove in the least beneficial to the general cause, the writer will feel highly satisfied. He only laments, that his obscurity and feeble abilities disqualify him from taking so active a part as he could wish in so noble and useful an undertaking.

He flatters himself, however, that his errors will be candidly passed over, when he declares, he had little leisure time to spend in this pleasant employment, except in the evenings, amidst a noisy groupe of young children, in which situation correctness was impossible. He has not vanity enough to suppose his name can be of the smallest consequence, therefore begs leave to subscribe himself, with the utmost deference and respect, to the patriotic President of the Board,

A Yorkshire Farmer.

ADDENDA TO CHAP. IV.

THE author of the "The Prefent State of Hufbandry of Great Britain" (Mr Donaldfon of Dundee) has in p. 240, vol. 4. of that work, rather ftepped out of his road, to make a thruft at us, for what we have faid on the important article of reftrictive covenants. According to him, "we have, by our attempt to get our brethren in the Weft Riding relieved from the improper covenants frequently engroffed in their leafes, not only materially injured the caufe we meant to ferve, but alfo the caufe of agriculture in general." This is certainly a weighty charge, and of fuch a nature as can only be juftified by proofs of the ftrongeft kind.

But what facts does Mr Donaldfon bring forward in fupport of his charge? What inftances does he give of the injuries done by us to our brethren, or to the general caufe of agriculture? Has the landed intereft of Yorkfhire relaxed the cuftomary covenants, or are the farmers of that diftrict allowed difcretionary powers in the management of their farms? As Mr Donaldfon refrains from ftating a fingle fact to fubftantiate his accufation; as he does not condefcend upon a fingle injurious confequence, either to individuals or to the public, from our Report; as the nature of the connection betwixt landlord and tenant, continues upon its former footing, we may, without prefumption, confider the attack he makes upon us, as wanton and unfounded.

If we had wished to answer this unprovoked attack, in the manner it deserved, we could easily have refuted Mr Donaldson with words, extracted from his own book. As for instance, he says, p. 232, same volume, " That the Legislature, the Board of Agriculture, and the proprietors of the country, may adopt what measures in their wisdom may appear proper to improve the national territory; but unless they go to the root of the evil; unless they adopt such measures, as will tend to place the British farmer in a more comfortable situation, and more on a footing with merchants and manufacturers, the object will not be attained." Now what do we say more? Is not the whole scope and intention of our Report to place the farmer in the comfortable and independent state recommended by Mr Donaldson? The merchant and manufacturer certainly lies under no restriction in the management of their affairs. They may carry on trade in the manner which will return them the greatest profit, or manufacture such commodities as the market requires. We contended for similar liberty being given to the farmer, and decidedly join Mr Donaldson in thinking that unless it is granted, the national territory will not be improved to its greatest height.

As Mr Donaldson's book contains a great deal of sound practical information, we cannot but lament that he should countenance a system which in a great measure restrains the farmer from putting it into practise. It is almost as absurd to expect improvements from a person whose powers are limited by restrictive covenants, as to believe, that a horse could gallop whose legs are tyed together. Perhaps the line of business hitherto followed by Mr Donaldson, may have biassed his mind upon this occasion, and influenced him to enter the lists in defence of arbitary or restrictive covenants. We are firmly convinced, that these restrictions have hitherto been of the greateſt prejudice to the extention of improvements, and that while they are continued, they will operate in the ſame injurious way.

We perfectly agree with Mr Donaldſon, reſpecting the propriety of cropping in a judicious manner, but differ widely whether the landlord or tenant is beſt qualified to judge upon the rotations to be practiſed. In every other line of buſineſs, a queſtion of this kind would be decided at once in favour of the operative perſon, and we have good reaſon to believe, that the majority of practical agriculturiſts will give a ſimilar verdict upon this occaſion.

Mr Holt, the Lancaſhire ſurveyor, in the reprinted Report of that county, alſo *attempted* to make a ſtroke at our Survey on account of what we ſaid in favour of leaſes. It was indeed but a feeble one, and might have been parried with words borrowed from his own work. In the 2d paragraph of page 25, the ſentiments ſtated by him, upon leaſes and covenants, are preciſely of the ſame nature as thoſe we ſubmitted to the Board's conſideration, and yet he pretends to tax us with abſurdity, becauſe we recommended free and open leaſes. We decline noticing the matter further; for an author who is not conſiſtent with himſelf does not deſerve to be noticed.

ADDENDA TO CHAP. V.

Owing to Mr Meikle's advanced ſtate of life, and other cauſes, his friends lately judged it expedient that a limited aſſignation of his patent, for the conſtruction and erection of threſhing machines ſhould be granted in favours of a deſerving and ingenious young man, Mr Thomas Wigfull at Aldwark near Rotherham, in the Weſt Riding. We underſtand, ſince that tranſaction took place, a great number of machines have been erected by him for the gentlemen and farmers in the Weſt Riding. From an intimate knowledge of his abilities we venture to recommend Mr Wigfull, as highly meriting every mark of public favour; and, from the aſſiſtance which this gentleman has received, and will continue to receive from Mr Meikle, we preſume the machines erected by him, will be found capable of executing work in the completeſt manner.

Mr Wigfull's aſſignation being limited to 20 counties, viz. Durham, York, Cheſter, Lancaſter, Stafford, Derby, Nottingham, Lincoln, Cambridge, Norfolk, Suffolk, Eſſex, Middleſex, Hertford, Bedford, Northampton, Rutland, Leiceſter, Warwick, and Huntingdon; Mr Meikle is ready to treat with perſons properly qualified in the remaining counties of England and Wales, for a further aſſignation of his patent right. We underſtand that he will aſſign for one or more counties, or the whole of them, as is moſt agreeable to the public; and it gives us pleafure to learn, that he has recently entered into a compromife with fome perfons who had, from ignorance, or other caufes, incroached upon his patent right. From an intimate knowledge of Mr Meikle's difpofition, we are certain that fuch compromifes will be made by him, upon the moft liberal terms, and we hope that an ingenious mechanic, who has benefited the intereft of agriculture fo much, by inventing the threfhing machine, will not in future be deprived of his reward.

APPENDIX.

APPENDIX.

No I.

EXTRACTS FROM THE JOURNAL KEPT DURING THE SURVEY.

WHILE employed in ſurveying the huſbandry of the Weſt Riding, a Journal was regularly kept of the information received at the different places we viſited an abridgement of which was given in the firſt edition of the Survey. When the extent of the Weſt-Riding, the varieties of ſoil, and the different modes of cultivation, are conſidered, it is hoped that a ſelection of the principal articles contained in the Journal will be uſeful and entertaining.

We commenced our ſurvey at Boroughbridge, on the 24th October 1793, which place is a market town, ſituated upon the river Eure, on the great north road from Edinburgh to London, and diſtant about twenty miles from York. The ground in its vicinity is of a good quality, being moſtly a deep loam, and a conſiderable part of it kept in graſs. Where tillage is practiſed, the ufual courfe upon light foils, is turnips, barley, clover, and wheat; and, upon heavy foils, wheat and beans, or wheat and oats: it being the cuftom of the country only to take two crops for one fallow, which is undoubtedly a great lofs both to proprietors and tenants. If land is fufficiently cleaned when under the operation of fallowing, and properly cropped afterwards, it can by no means be in fuch a fituation as to require that the produce of one year fhould be facrificed to afford the means of a third crop.

Copgrove is the feat of Henry Duncombe, Efq; who keeps a confiderable quantity of his eftate under his own management. The foil is light, and excellently fitted for raifing carrots, turnips, and other drilled crops. The inclofures are well laid out, the fences kept in good order, and the pafture grafs particularly fine.

The ground about Knarefborough is moftly kept in grafs, and employed in feeding milch cows. Thefe are generally of the Holdernefs breed, and are excellent milkers; but a little farther weftward, the Craven, or long-horned breed prevails. It gave us furprize to learn, that lime is applied in this neighbourhood in fuch fmall quantities, no more than 64 bufhels being ufed for a ftatute acre, and often only half that quantity.

The foil and climate vary according to fituation, expofure, vicinity to rivers and towns; as alfo to the quantity of lime, compofts, and other manures that are ufed. Farms are in general fmall, and divided nearly into equal portions of arable and grafs; all kinds of clover and grafs feeds are fown. A mixed ftock of horfes, cattle, and fheep, are kept upon the paftures, but the breeds are by no means properly attended to, except by fome particular perfons. Very little land is watered, tho' many fituations would admit of it. Fallowing practifed invariably upon ftrong foils, and even upon all ſuch as are not friendly to the turnip huſbandry. Turnips, when eat by ſheep, ſeldom fail to improve the ground, and ſecure a ſucceſſion of good crops; but red-clover, when too often ſown, is found not ſo good a ſucceſſion as trefoil and white-clover, or even beans, occaſionally introduced in its ſtead. Lime is principally uſed as a manure, when ſtable dung cannot be obtained. Compoſt dunghills are now more attended to than formerly, but not ſo much as they ought to be. Many of the common fields are incloſed annually, under particular acts of parliament, and by them population has been greatly increaſed. The extent of waſte land is very great in this Wapentake, and principally depaſtured by half ſtarved ſheep, horſes, and young cattle; it may be improved in various ways, as the commons in this diſtrict differ much in ſoil, expoſure, and other circumſtances. The land is not ſo much drained as it ought to be, the drains are moſtly filled with ſtones, and covered; very few filled with wood or *ſtraw*. If the ſoil be ſound and ſtrong, it is common to turn the firſt ſod, with the graſs ſide downwards, letting it reſt for ſupport on a piece of the bottom of the drain on each ſide not thrown out; this is called a ſhoulder drain, and in ſtrong land anſwers well, and is done cheap. Paring and burning is practiſed, and found to anſwer well in all ſour ruſhy land, and is done by men, with a puſh or breaſt ſpade. In ſome parts of the country wood abounds, and where it is attended to, thrives well. The roads are in general good; thoſe are beſt which are made wide, not too much raiſed in the middle, and the ſtones broke ſmall, by which means they unite and bed firmer. The farm-houſes and offices, when made in conſequence of new incloſures, are uſually placed near the centre of the farm, and are well conſtructed. Few leaſes are granted, which is to be lamented, as it can never be expected that improvements will be made, where the tenant has no certainty of reaping the benefit of them. The people certainly have a great turn to improvements, and were reasonable leases given, would make a rapid progress therein. The intermixture of property is considered as a great obstacle to improvement; and we were informed, that if a general inclosure bill could be obtained, and tithes commuted, it would obviate a great many of the present impediments, and contribute more to extensive and general improvement than any other measure.

The forest of Knaresborough, adjoining to this place, and consisting of 33,000 acres, was divided in the year 1770; and an account of the difficulties which occurred in accomplishing a division, will be found in the preceding part of this work. For this, as well as various other important informations, we were under great obligations to a gentleman of this place.

Visited Harrowgate; country about that place wild and uncultivated; in the division of Knaresborough forest, a tract of land, about 200 acres, was set apart for the use of the company who resort there.

At Ripley, we learned that the greatest obstacle to improvements was obliging tenants to keep their land constantly in a state of pasturage. From hence to Paitley Bridge, the face of the country alters exceedingly—a great deal of waste land, the road unequal, and shaded with trees. There is a fine valley of land called Nidderdale, in which the river Nid flows; but the higher ground appeared mostly to be in a state of waste. The Dale is very populous, and the inhabitants are much engaged in the linen manufacture. They generally bleach the yarn before it is wove, which we were told contributed to strengthen the cloth. A good deal of butter is likewise salted here for the London market, and a cow pastured upon the low grounds, is computed to yield 3 firkins of 56lb. each, during the season. A number of hogs are also fed upon oat meal, and sold to manufacturers in Lancashire.

In the neighbourhood of Paitley Bridge, there are a considerable number of lead mines. The land, as we proceeded to Grassington, was of inferior quality; a great part of it common or waste, and pastured with sheep of a bad sort, and in as bad condition. About Grassigton the soil turns better, and the low grounds are all inclosed. Oats are the chief grain sown here; but almost the whole of this neighbourhood is kept in grass, and employed in feeding cattle and sheep for the Skipton market.

Leaving Grassington we passed through a wide range of uncultivated moors, and arrived at Settle. At this place we saw the finest grass we ever viewed. Indeed the richness of the soil is hardly credible to those who have not seen it, and the possessors were unanimously of opinion, that it is of greater value to them when kept in grass, than when cultivated by the plough.

The nature of the soil in the neighbourhood of Settle, is what is called a hazle mould, incumbent upon a dry bottom. The farms are generally small, and the occupiers seldom have leases. Great part of the higher grounds are still common, and consequently unimproved: They are pastured with sheep and Scots cattle, which are afterwards fed off upon the lower grounds. The sheep bred here are called the Malham breed, and we received favourable accounts of them. Considering the great quantity of waste ground, it is surprising the proprietors have not turned their attention more to planting, as we received great complaints of the scarcity of wood. Coals are likewise scarce, which it was thought might be remedied, if proprietors were disposed to hold out rewards or favourable leases to these who discovered them.

At Settle we had an opportunity of seeing a great show of fat cattle of the country breed. They were all long horned, and seemed in shape, skin, and other circumstances, to be nearly the same as the Irish breed. We learned, that of late there had not been the same attention paid as formerly to keep the breed pure, by selecting proper bulls. Be this as it may, the long horned breed of cattle, which prevails over the western part of the island, from the thickness of their skin, and the hardness of their constitutions, are much better calculated to undergo the vicissitudes of this climate, than the short horned breed of the eastern coasts.

Left Settle and proceeded to Ingleton. The land all inclosed, and almost wholly in grass of the richest quality. No turnips to be seen since we left Pately Bridge, and hardly a stack of corn. In short, from the plenty of grass, and scarcity of corn fields, we were ready to conjecture that the inhabitants of this part of Yorkshire lived upon butcher meat altogether.

Leaving Ingleton, we proceeded for Dent Dale, the most western extremity of the county.

Upon the road we called upon Bryan Waller, Esq; at Maisongill, from whom we had the following accounts of the husbandry in his neighbourhood:

Soil a strong loam, and from the wetness of the climate, unfit for ploughing—generally possessed by small proprietors, and partly set upon leases of 3, 7, and 9 years. Land set here by the customary acre, 3 acres of this measure being equal to 5 statute acres. Small tithes paid in kind, and a modus taken in lieu of hay. The farmer allowed to plow but a small part of the land, often but eight acres where he possesses a hundred. Ploughing more practised formerly, but breeding and feeding cattle is now thought more profitable. No turnips. Cattle fed in the house during the winter months upon hay, which renders beef very high in the spring. A number of Scotch cattle wintered upon the pastures, which are disposed of by Midsummer; the commons are stocked with Scotch sheep, the large breed being thought above the pasture. Cattle that are bred here are all of the long horned kind. No land watered—thinks it would be hurtful in this cold country. Very little fallow, and no attention paid to the plough. Lime applied to the pasture grass, and mixed with earth and cow dung—the remainder of the dung laid upon the fields that have been cut for hay. Plough wrought with three horses, often four, and all yoked in a line. Land all inclosed, except the commons, partly with hedges, and partly with stone walls. Inclosing has increased rents greatly. A great deal of waste land in the moors; which he cannot say is improvable, as planting is not found to answer. Wages high.—Labourers 1s. 6d. per day and victuals, during hay time and harvest. Some of the lands are drained—shoulder drains have been found to answer upon mossy soil, where it is improper to put stones; but in general all drains are built with walls, and covered with flags. Thinks paring and burning not good farming. Wood very scarce in this part of the country. Farm houses rather stand too much in the villages, and therefore inconvenient. Some cotton mills which employ a good number of hands—no other manufactures. Does not think the people trouble their heads much about improvements, and thinks the present stock of sheep well adapted to the soil and climate.

Continued our journey to Dent.—A great deal of good land, but the general quality of the soil thin, and a moist bottom. Learned that there was a considerable quantity of butter salted in this tract, and disposed of at Skipton.

Arrived at Dent after a tedious and disagreeable journey, having, in the course of it, passed through a small part of Lancashire, and travelled about eight miles in the county of Westmoreland.

We entered Dent Dale from the west, and proceeded down the Dale to the town of Dent, which is nearly in the centre. This Dale is entirely surrounded with high mountains, and has only one opening from the west, where a carriage can enter with safety. It is about 12 miles in length, and from one and a half to two miles in breadth. The whole Dale is enclosed; and, viewed from the higher grounds, presents the picture of a terrestrial paradise.

At Dent we received the following information relative to the state of the Dale—

Estates are small, and chiefly in the natural possession of the proprietors. Inclosures small, and mostly grass. No farms above L. 50 a-year, and none but yearly leases granted. Sheep mostly from Scotland. Few cattle are fed for the butcher, but a great number of milch cows are kept, and large quantities of butter and cheese produced. The hills in the neighbourhood of the Dale, are all common, and dividing them among the different proprietors, it is supposed, would be attended with beneficial consequences. A considerable quantity of stockings wrought by women upon wires, which are disposed of at Kendal. Very few turnips cultivated, hay being the chief dependence in winter. Small tithes only drawn in kind, and a modus taken in lieu of the great ones.

Returned from Dent to Ingleton, where we met, agreeable to appointment, with Mr Ellershaw, of Chappel le Dale, about four miles from this place. Mr Ellershaw gave us the first account of watering land, which is done by him, and several of his neighbours, to great advantage: he floats it early in the spring, which not only rots the moss, but enriches the land considerably. The commons here are all flinted, every man who enjoys a privilege being restricted to the quantity of stock he is to put on them. There is not much land limed in the neighbourhood, and what is done, is applied very sparingly. Few or no leases granted; and these are, of short duration. Tithes drawn in kind; but Mr Ellershaw thinks it would be for the peace and interest of the community to have them valued. No turnips raised. Sheep generally of the Scotch kind. Wool sold at 6s. 3d. per stone this season. Some stockings knit for the Kendal market.

At Gargrave, half way between Settle and Skipton, we saw most excellent fields of grass. It is impossible to say what sorts of seeds had been sown, or whether any had been sown at all; they seemed a mixture of all sorts of hay seeds, but richer grass cannot grow.

Arrived at Skipton. This place which stands in the middle of the district of Craven, is for distinction usually called Skipton in Craven.

At Skipton there is a large house employed in sorting and combing wool. About 3000 packs are bought each season from Lincoln, Nottingham, Leicester, and Rutland shires. After it is sorted and combed, it is spun at the companies mills, at Linton and Addingham in the neighbourhood, and made into stuffs, viz. shalloons, callimancoes, and all sorts of double goods. The noyles from the combing are used for the Dewsbury and Rochdale trade.

The proprietors in the vale are, the Duke of Devonshire, Lord Thanet, and a number of small freeholders. Farms of different sizes; but the majority rather small. Soil deep and rich. The whole vale almost in grass, being from the wetness of the climate accounted unfit for corn. What land is ploughed, is upon the higher ground, and oats the principal crop. Few or no turnips cultivated. All the vale is inclosed. Inclosures small. Very little wood but a great part of the moors might be planted to advantage*. Provisions high, beef being at this time 4d. and often 5d. and 5½d. Corn brought here from Richmond in the North Riding. Roads good. Farm houses in general well situated. Lord Thanet's estate upon lease of 14 years. Duke of Devonshire grants none. Lord Thanet formerly granted leases for 21 years, and the estate was much improved. Other estates where leases have not been granted are not half so much improved. The covenants laid down by Lord Thanet are only to fallow, lime, and manage in a husbandlike manner. No manufactures except some cotton mills which have done no harm to the agriculture of the country. Grass lands in the vale set from 40s. to 50s. per acre, and some at L. 3. Plough yoked with three horses, no oxen used.

The management from Paitley-bridge to the western extremity of the county, is almost uniformly the same, and grass the sole object. The people unanimously think that corn will not pay so much rent as grass, therefore raise very little, except upon the higher grounds; and at the same time lay all their manure upon the rich, fertile grass fields in the vale. By this means they are reduced to the absolute necessity of purchasing corn, at an advanced price, from other places, where more attention is paid to the cultivating it. From what we could learn, a great deal more corn was formerly raised than now;

* Not a doubt of it. Scarcely a bleak hill in the island where wood of the proper sort will not thrive. Many a spot is condemned by planters for want of ascertaining in a small nursery on the place, what kind of trees will suit the soil and climate, previously to the formation of any plantation.

Mr Payne, Frickley.

which is evident from tithes having decreased four-fifths in value within these thirty or forty years.

We learned from the Rev Mr Wethnell, that the husbandry of Keighly is much in the same style as here, only rather more corn raised, and that the moors and high grounds are used for breeding cattle.

From Skipton proceeded down Wharfdale to Otley. For the first four miles the soil is barren moor, and perfectly unimproveable, unless planting will answer, which, from the highness of the situation, is very doubtful. At Addingham, the soil turns good, and the whole way to Otley remarkably rich. At Sir James Ibbetson's, at Denton, there are fine haughs of grass, and the inclosures larger than we have seen in Yorkshire. Saw some corn fields upon the road, but not in good order; and a few fields of fallow, not half wrought. Observed some turnips, the first we have met with for a long time. Examined a plough; the first we have seen these four days, and it appeared to be of the Dutch or Rotherham kind, but of wretched construction. The field it was lying in was full of quickens, provincially *whickens*. The land almost wholly inclosed.

At Otley the soil is good and the climate dry. Some large proprietors, but a greater number of small ones. Farms chiefly small, few above L. 50 rent. Land mostly employed in pasture, and sown with white clover and hay seeds. Little land watered; but understood some people have done it to good effect. When land is broke up from grass, three crops are taken, and then a fallow. Few turnips are cultivated. Some lime used. Harvest early. Land all inclosed, and thought much more valuable than when open field. Wood scarcer than formerly, but a great deal of the waste land might be planted to advantage. Tithes compounded at 5s. and

6s. per acre. Rent of land here 40s. per acre, besides public burthens. Want of leases greatly complained of. Some cotton mills, which have done good, by employing young people. A common lately divided in the neighbourhood, which has turned out well.

Arrived at Leeds. Leeds is situated on the river Aire. It is a very ancient, and populous town, and was of considerable repute during the Saxon government. The woollen manufactory has flourished here for several ages, which has both enriched the inhabitants, and increased the value of all the land in the neighbourhood.

The following is the most accurate accounts we could procure of the state of husbandry near Leeds:

The soil variable—a great part of it good, generally loam upon a clay bottom. Climate dry. Land possessed by small proprietors, and mostly occupied by manufacturers: a few of what are here called large farmers, having from 100 to 150 acres of land. Land employed partly in pasture, partly in meadow, and a proportion in tillage, but ought to be all in grass upon account of the great demand from Leeds for milk. Some clover and rye-grass sown. The stock kept upon the pastures are cows and horses belonging to the manufacturers. Part of the land watered and turns out well. Grains cultivated are wheat, barley, oats, and beans; also some rape, and turnips, which are generally sown broadcast. A few beans are drilled. Fallowing much practised. Large quantities of potatoes raised, and a great demand for them. Much lime is used, and both grass, and fallow dunged. An excellent manure is got from the sizing boilers' waste, which is the bones and remains of sheep feet, cows feet, and sloughs of horns. Horses only used—Seed time, and harvest early. Land mostly inclosed, and rents greatly raised thereby. Inclosures from 5 to 8 acres, and the smallest ones most valuable, being poſſeſſed by clothiers, who have no uſe for large ones. Incloſing in a manufacturing county muſt increaſe population. Very little waſte land but what might be improved by dividing and incloſing. Wages: Maſons 15s. to 18s. per week. Carpenters the ſame. Mill-wrights, 18s. to 21s. and day-labourers 9s. to 12s. Journeymen clothiers from 9s. to 15s. per week.* Ploughmen L. 12 per annum, with victuals and beer. Very little paring and burning, unleſs where commons are newly divided—the expence from 19s. to 27s. per acre. Not much wood, as land can be uſed to far greater advantage otherwiſe. Proviſions high —Beef 4½d. and 5d. and much higher in ſpring. Roads in general but tolerable, owing to their being let to undertakers, who neglect them. Houſes for manufactures well conſtructed; and a great many more wanted. Few leaſes—when granted, their duration from 3 to 15 years: The nature of the covenants in them is, that the tenant pays all taxes, keeps all in repair, is bound not to break up any graſs land, under penalties, that run from L. 5 to L. 20 per acre, and to have at leaſt two thirds of the farm in graſs; upon the tillage part he muſt not take above 3 crops without fallowing, and all the fallows muſt be limed.

Broad cloth and other kinds of woollen goods are manufactured here, which has greatly increaſed rents. There are no agricultural ſocieties, but the people have a great turn for improvements—the expence is not regarded.

* Yet two years ſince the cry was, among ſome of the rich; "there wants a war to reduce wages," horrid expedient, horrid motive! Who is ſo *well* entitled to a comfortable maintenance as the labouring clothier, from the fruits of whoſe toils the merchants, &c. attain their immenſe fortunes? But it is plain, there are ſome *callous* ſouls, who are never happy but when the poor are miſerable. I had no idea at the time that 15s. per week was the *maximum* of wages ſo grudgingly paid. *W. Payne, Eſq.*

Several inclosure bills passed for moors in the neighbourhood, which have produced the most beneficial consequences. Tithes, both small and great, drawn in kind; but the general opinion is, that a compensation in money, in place of them, would operate as a great encouragement to improvements. It was also the opinion of our informers, that a general inclosure bill, upon proper principles, would be of great public utility; as by it, they said, much expence would be saved to individuals,* houses would be provided for manufacturers, and the people prevented from emigrating.

Left Leeds, and proceeded to Bradford. Rather more corn land than we have as yet seen during the survey, but all in bad order. A good deal of oak wood at Kirkstall Abbey, about 3 miles from Leeds. Observed a plough at work, and drawn by 4 stout horses all in a line. The plough of a very indifferent construction, and taking a very ebb furrow, not the depth of what 2 horses will do when properly yoked abreast—the land very much damaged by the large sweep the horses took when turning. Saw another plough upon a soft moor going with 3 horses—a very ebb furrow but the straightest we have as yet seen in Yorkshire. Indeed in those parts of the country we have hitherto surveyed, ploughs are so scarce, that they may almost, like horses at Venice, be shewn as a curiosity.

The nature of the soil in the neighbourhood of Brad-

* But what would become of the *poor* but *honest* attorney, officers of parliament, and a long train of &c. &c. who obtain a *decent* livelihood from the *trifling* fees of every individual inclosure bill—all these of infinite use to the community, and must be encouraged whether the wastes be inclosed or not. The waste lands, in the dribbling difficult way they are at present inclosed, will cost the country upwards of twenty millions to these gentry &c. which on a *general* inclosure bill would be done for less than one. *W. Payne, Esq.*

ford is various, some parts being rich loam, and others of a cold watery quality. Climate healthful. Land is possessed by small proprietors, and occupied by small farmers and manufacturers. It is almost all in grass, and the seeds sown are mostly those called, natural hay-seeds. Cows are the principal stock that is kept. Where the land is in tillage, wheat and beans are sown in small quantities, but oats are the principal crop. Some good farmers adopt the modern rotation of turnips, barley, clover, and wheat. Fallowing is practised, but often in a very slovenly manner, and the rotation in that case is, wheat, oats and oats; or wheat, beans, and oats. The country is all inclosed; inclosures small, few exceeding 6 acres, and by them the country has both been enriched and the land improved. Labourers wages 9s. per week. Ploughmen L 12 per annum, with victuals, lodging, and washing. Paring and burning only practised where heath ground is broke up. Few leases are granted, those that are, generally for 11 years, and the covenants are, to lime all the fallows; not to take more crops than 3; to keep the premises in repair; not to sell hay, straw, or manure, provincially, *tillage*; and not to assign. No practises can be pointed out here, that would be of advantage in other districts, the inhabitant having both their minds and capitals fixed upon trade.

Arrived at Halifax—the whole country from Bradford to this place, being almost a continued village, roads bad ever since we left Leeds, and materials very scarce. Observed most of the roads provided with a foot-path, paved with free stones, which is a most useful measure; but, strange to tell, every person upon horseback uses the foot paths.

Halifax is a large and populous town, and is situated upon the river Calder. It stands upon a gentle descent from east to west, which makes it pleasant and convepient. The houses are, in general, built of brick, though free-stone appears to abound in the neighbourhood; and as little attention appears to have been paid to the laying out of the town at first, the streets and buildings are rather irregular and confused. The woollen manufacture has here flourished for near three centuries; and though the soil in the neighbourhood has been originally barren, and probably for that reason, was chosen as a proper place for carying on manufactures, yet the industry of the people has been so great, as to improve almost every spot near the place, thereby making good the old proverb, that a barren soil is an excellent whetstone to promote industry.

The parish of Halifax is in the wapentake of Morley, and consists of 26 townships or hamlets. The parish is of great extent, and supposed equal in size to the whole county of Rutland. It is about 17 miles in length, and 11 in breadth. From this extent it clearly appears that the ground must have been a barren waste, and the population of small amount, when the parish was formed. Several parts of the parish afford coals, which are absolutely necessary for carrying on that extensive trade, for which this neighbourhood is famous. The air is good, and chiefly blows from the west and south-west, and often attended with heavy falls of rain; but, as there is but little level land in the parish, the rain which falls soon runs off, and of consequence the country is clean and dry, which contributes both to the ease and health of the inhabitants.

We observed some fences of a very uncommon kind in this neighborhood. Large flag stones of 3 feet height, set upon their end, are fastened in the ground, which make a fence both complete and agreeable. We cannot speak to the expence, but as stones of that kind are here in plenty; we suppose a fence of this kind will be comparatively cheap.

Waited upon William Walker, Esq; at Crow-nest near Halifax, and examined his improvements, which are executed with singular taste and ingenuity. Mr Walker waters his ground with great success, which is all laid off with much attention for that purpose. All his inclosures are in perfect order, and his farm offices are in the neatest condition.

Mr Walker was so kind as to favour us with the following account of the husbandry of the parish of Halifax:

The soil varies much, but in general is naturally poor. Proprietors both large and small. Farms mostly small, and occupied by manufacturers, for the conveniency of keeping a cow or two for the use of their families, and horses for conveying their goods to the mill and to the markets. The land is principally in meadow pasture grass, and is sown with natural hay-seeds, rib-grass, and rye-grass; and where it is not used by the manufacturer, as mentioned above, it is pastured with a mixed stock of horned cattle and sheep. Great advantages are found to result from overflowing the meadows at proper seasons, and particularly in time of floods. Land is generally fallowed after the third crop. Sometimes turnips are taken upon the fallow, then barley, clover, wheat, or oats. A small quantity of wheat is sown, and very few beans. The lands, except the heathy moors, are mostly inclosed; but there are doubts, whether any advantages at all have resulted from inclosing the waste lands in this parish. The size of inclosures are in general from 2 to 4 acres. Inclosing in this parish has certainly had no tendency to decrease population. The extent of waste ground, if we include the heath, can scarcely be guessed at. It is, however, very considerable, and there is some worth the expence of inclosing for cultivation: at any rate, it is worthy of confideration, whether it is not a defirable object, that each freeholder's property be afcertained, that fuch as are inclined to improvement, may do fo by planting, or otherwife. Wages high; hufbandmen get from 18d. to 20d. per day; in time of harveft 2s. Great attention is paid to draining, which is done in a complete manner with ftones. A very inconfiderable quantity of wood-land in this parifh. Price of provifions, butter 12d beef 3½d. and 4½d; mutton 4½d. and 5d.; veal 4d. and 4½d. The roads are very bad. The houfes and offices are built for the accommodation of the manufacturer, not of the farmer. Leafes are granted for various terms, from 7 to 21 years; but very frequently no leafes at all are granted. The principle manufacture here is woollen and worfted goods, and fome cottons. Manufactures are the grand object of perfons of all defcriptions, and the land is divided into fmall farms, in aid of the manufacturer. There are very few who attend, in any degree, to the cultivation and improvement of the ground, which is regarded only as a fecondary object.

Set off for Wakefield. The foil appeared thin for a confiderable part of the way, and rather of an inferior quality. At Dewfbury the ground turned better, and a number of fine fields appeared upon the banks of the Calder below that place. The road from Halifax to Wakefield was in moft fhocking condition, and the heavieft ftage we have travelled. Obferved the materials are of bad quality, and that to render them harder, a great part of them are burnt before they are laid on the road; alfo that clay was burnt into a kind of brick, and ufed likewife for repairing the roads. Want of proper materials is a local difadvantage, for which the road furveyors can never be blamed. They feemed however, to us, to be carrying on repairs upon bad principles: inſtead of filling up the old ruts, which were very deep, and levelling the ſurface, a new covering was laid on indiſcriminately, which will never bed firmly, or conſolidate in any ſituation. Beſides, the repairs were carrying on at an improper ſeaſon; for the roads appeared to receive conſiderable damage from driving the materials.

Saw three large ſtrong horſes this afternoon, drawing a light break harrow, which might have been eaſily worked with two. The horſes in this part of the country go uniformly in a line, and ſeem much ſtronger than any we ſaw in the northern parts of the Riding.

Wakefield is a large well-built market town, and poſſeſſes a conſiderable ſhare of the cloathing trade. It is very populous, and has two market days weekly, at which great quantities of cloth, wool, corn, and proviſions of all kinds are ſold. It ſtands upon the river Calder, which by an act of Parliament in 1698 was made navigable as far as this place. A canal is, at this preſent time, making from hence to Barnſley.

From Wakefield to Pontefract, the ſoil is much drier, and corn fields more numerous. Paſſed a large common field, which appeared in very bad order. Arrived at Pontefract, and met with a number of intelligent farmers, from whom we received much information. They all concurred in one ſentiment with regard to tithes, viz. that it would be a material encouragement to improvements if they were commuted; alſo that every common field in the kingdom ought to be divided.

Waited upon Mr Green at Cridling Park, near Ferrybridge. Mr Green rents this farm from one of the colleges at Cambridge. Is a complete farmer, and keeps his land in good order, but is abſurdly reſtricted by his leaſe from breaking up old graſs.

Proceeded ſouthward for Mr Gill's at Notton. The

lands upon the road are of good quality, and well farmed. Fallows clean. Saw some very large fields of wheat making a vigerous appearance.

Information from Mr Gill,

Soil generally of good quality, part of it gravel, the rest clay upon a wet bottom. About two-thirds of the ground kept in tillage, and one-third in pasture. Red and white clover sown with rye-grass. Breeds a few horses, and feeds both cattle and sheep. Rotation—fallow, upon which turnips are taken, barley, clover, and wheat, sometimes oats. Uses a good deal of lime, but applies only 30 bushels to the acre. Brings great quantities of bones from Sheffield, which is at 20 miles distance, and lays on 50 bushels per acre; costs from 15d. to 18d. per bushel, besides carriage. Plough of the Dutch kind, and wrought mostly with two horses abreast, but sometimes with four in strong land. Carts of the ordinary construction of the country, and drawn by three horses. Land all inclosed, which Mr Gill thinks of great advantage. Size of inclosures from 2 to 14 acres. Thinks small inclosures very hurtful. There are several common fields in the neighbourhood, which should be divided and inclosed; very little waste land; wages of a ploughman, L. 11 per annum, and victuals. A good deal of land is drained, big stones being set in the bottom of the drain, leaning towards one another, and filled up with small stones. Paring and burning practised here; but the landlord's consent must be got—expence 20s. per acre. Roads generally good and well managed, but materials bad. Funds are 6 days labour of a team for L. 50 rent, and 9d. per pound assessment upon the rent. No leases granted, which he thinks retards improvements. Tithes paid for in money, at the rate of 6s. or 7s. per acre. Sheep in this neighbourhood are either of the Scotch kind, or purchased at Peniston, from the moors in the western parts of the county: the wool of the former sells for about 8d. per lb. the other 9d. People have a turn for improvements, and know no obstacles but the want of leases, and payment of tithes.

Arrived at Bretton hall, the seat of Mr Beaumont, and experienced the greatest attention from that gentleman. He was at the trouble personally to shew us a part of his large estate, which is farmed in as complete a style as any in Yorkshire. Saw very fine broadcast turnips at Mr Brook's, one of Mr Beaumont's tenants: they were remarkably clean, a thing rather uncommon in this country. Were introduced to several of the tenants, whom we found sensible, industrious men. They were busy sowing their clover leas with wheat. Their young grasses were making a most vigorous, close, and equal appearance. Mr Beaumont has a good deal of wood upon his estate, which is very thriving and profitable. Farm-houses and offices are excellently constructed, and well situated. Understood the late Sir Thomas Blacket, Mr Beaumont's father-in-law, was very attentive to these matters; and although he granted no leases, was otherwise a kind and indulgent master.

The soil here is variable, chiefly hazle kind of earth, mixed with clay and a loamy sand, both retentive of water. Some parts dry and sharp, well adapted for turnips, which are generally cultivated upon all the fallows, and eaten with sheep. Proprietors here, are Mr Beaumont, Mr Wentworth, Mr Stanhope, &c. Size of farms from 150 to 200 acres. Land chiefly in tillage; one-third only kept in pasture; several rotations of crops are practised.—1st, fallow, wheat, oats, and barley. 2d, turnips, barley, clover, and wheat. Often hay-seeds and white clover is sown with the barley, upon which sheep are pastured for two or three years. No land watered, but thought adviseable when opportunity allows. Manures used are dung, lime, rape dust, and lately a great deal of bones. Mr Hague, one of Mr Beaumont's tenants, says bones answer best on the turnip land, 100 bushels of bone, and four loads of dung, mixed with good earth, is laid upon a statute acre. Quantity of lime applied to the acre, generally about 50 bushels. Rape dust one chalder per acre, price L. 3 : 12s. besides carriage. Rape often sown for sheep feed, but not cultivated for seed; at least what is done, is in very small quantities. Carriages with broad wheels are used for the fields, and narrow wheels for the roads Ploughs used are of the Duch or Rotherham kind, yoked sometimes with two horses abreast, at other times three in a line. No oxen used; wheat sown from the beginning of October, to the end of November; sometimes to February, after turnips: but that season not approved of: spring corn in March and April. Harvest variable; generally commences about the 18th August, and over by Michaelmas. Land chiefly inclosed; inclosing of great advantage, and thought to be one-fourth more value than open field. Inclosures from two to twenty acres; average about ten acres; inclosing thought to increase population. A few common fields in the neighbourhood, and these thought to be under bad management; very little waste land: wages for ploughmen L. 12 per annum, with victuals, washing, and drink. In harvest, labourers 2s. per day, and 2s. 6d. with beer. Hours of work from six to six, with one hour allowed to dinner, and another for the two drinkings: in winter from light to dark. Draining a most necessary article of improvement, and great attention bestowed on it; two stones being set up leaning on each other, and the drain filled up with small stones. Paring and burning practised, but not thought good farming; expence when done 21s. per acre, with beer, which makes it equal to 24s. A good deal of oak and ash wood in the neighbourhood, generally cut once in 21 years, a regular portion being done annually, sometimes sells so high as L. 80 per acre. Provisions—beef from 3 l. to 5 l.; mutton 4½ l.; butter 11d. and 1s. all avoirdupoics weight. Roads in this township good, but bad in many others; supported by the statute work of six days labour of a team for every L. 50 rent, and 6d. a pound assessment; but this rate may be raised higher by the justices, if they see necessary. Farm houses and offices good, and well constructed for serving the purposes of husbandry. Some manufactures creeping in, which are raising wages. The people here have a great turn to improvements, and have no obstacle but want of leases; which, from the kindness of their landlord, is little felt. There is no restriction upon their management that is hurtful, but one, which prevents them from breaking up their grass land that has lain six years, without the landlord's consent.

Mr Slinger at Emly Woodhouse, upon Mr Beaumont's estate, practises the drill husbandry both for wheat, pease, and beans; but does not think drilling at all times and in every situation, adviseable. Mr Slinger uses a machine for threshing his corn, which he thinks does the work much better than is done by the flail: and these machines are particularly necessary here, since wages became so extravagantly high. No want of hands in this neighbourhood to cut the crop. In the year 1792, the scarcity was great in the East Riding, but felt here no farther than rising wages. An inclosing bill lately passed for dividing land near Wakefield; but as it is not yet put in execution, cannot say how it will operate. Mr Beaumont's tenants were unanimously of opinion, that all commons ought to be divided, as they know some that formerly carried only a few beggarly sheep, now improved into rich corn fields.

Arrived at Barnfley, which is fituated in the Wapentake of Staincrofs; it carries on a confiderable trade in wire, and has a manufactory for bleaching and weaving linen yarn, which is in a flourifhing ftate. There is a weekly market held held here, where corn and all forts of provifions are fold. It being market day when we were there, had an opportunity to fee the quality of the different grains. Wheat and barley good, but the oats very indifferent, which in general we found to be the cafe over all the Weft Riding.

The land to the fouthward of Barnfley of the fineft quality, being either a clay or a loam fit for turnips, and a great proportion of it kept in tillage. Mr Hemmingway, at Wombwell, gave us an account of his practice, which is very correct. He keeps about one-fourth of his farm in pafture, which is fown down with white clover and hay feeds; fometimes fows red clover by itfelf, paftures it in the fpring, and then lets the crop ftand for feed; fows white clover for the fame purpofe, and has often 6 bufhels red, and 4 bufhels white, per acre. If good in quality, a bufhel weighs 66 lb. Employs his pafture to fupport his farm flock, and in feeding ewes and lambs—ewes of the long woolled kind from Northumberland, and rams of the Bakewell breed. Does not water any land, but approves of it when fituation allows. Cultivates turnips in large quantities, fome of them drilled. Fallows every fourth year, and manures with dung, rape duft, and bones. Plough of the Dutch kind, and wrought with two horfes abreaft. Carts long in the body, and of the fame conftruction with the reft of the country. Land moftly inclofed—inclofures from 5 to 15 acres. Does not think inclofing can ever decreafe population. Pares and burns old grafs land; expence 21s. per acre. Pays great attention to draining—makes the drains 2 feet deep, 18 inches wide at top, and 12 at bottom, and fills them with stones. Roads very bad; and materials scarce. Few leases granted, which he thinks a bad plan.

From Barnsley to Peniston the country falls off, being of a moorish soil near the latter place. A market for sheep is held at Peniston, and large quantites of those that go by that name, are sold weekly. They are bred on the moors to the westward of Peniston, and on those of Cheshire and Derbyshire—prices at present low, and sale dull. The climate cold and backward to vegetation. Soil very variable, but mostly wet and spongy, and a great deal of moor carrying little but heath. Proprietors small, Mr Bosville of Gunthwaite, the representative of one of the oldest families in the county, being the only large one. Farms likewise small, except upon the moors. In the vicinity of the town about one half is ploughed, but in the moors there is little or no tillage at all. The stock is sheep and long horned cattle, of the Derbyshire breed, which are smaller than the Craven breed. Little grain is cultivated, except oats and a small quantity of wheat. Dung chiefly applied to the meadow land that has been cut for hay, and 2 chalders of lime per acre laid upon the fallows. Plough wrought with 4 horses, yoked in a line. Few oxen used. Seed time and harvest late, sometimes November before the harvest is concluded. Some land about the place inclosed, but to the westward it is all common moors; which ought at least to be divided, and every man's property laid by itself. A great deal of the land needs draining, but the proper method of doing it not well understood. Farmers generally debarred from pairing and burning, but thought a great means of improvement upon some lands. Few proprietors grant leases. The Rev. Mr Horsefall, in answer to this question said, if he was a farmer, he would lay out his money more frankly under the ſecurity of a leaſe, than if he had none. Many reſtrictions are in the leaſes, or yearly bargains. Some farmers thought to need them, but an active induſtrious man hurt by limitations.

Left Peniſton for Sheffield. Moſt of the way the ſoil indifferent. Saw ſome patches of turnips, but none of them good. Road to Sheffield high, and very unequal. Fine country to the northward, and abounding with oak-wood.

Sheffield is ſituated upon the river Don, and has long been a ſtaple place for cutlery ware of all kinds. It is a populous town, containing not leſs than 40,000 inhabitants. The lord of the manor is the Duke of Norfolk, who likewiſe poſſeſſes a large eſtate in this part of the Riding.

The ſoil in the neighbourhood of Sheffield is generally a hazle loam, well calculated for turnips. Climate middling. Average gage of rain 33 inches in a ſeaſon, which is about a medium betwixt what falls in Lancaſhire, and on the eaſt coaſt. Large proprietors are the Duke of Norfolk, Earl Fitzwilliam, and Counteſs of Bute; but there is a number of ſmall freeholders. Farms ſmall in the neighbourhood of Sheffield, from 20 to 60 acres; and the Duke of Norfolk, upon his eſtate, is reducing their ſize as faſt as the leaſes expire, for the conveniency of the inhabitants. Near Sheffield, three-fourths of the land is in paſture; and, at a greater diſtance, about one half. Some red clover and rye graſs is ſown, but the general practice is to ſow white clover with hay-ſeeds. The paſture graſs is chiefly ſtocked with milch cows, and a few ſheep, which are moſtly of the Peniſton breed. Little land is watered, but approven of when it can be conveniently practiſed. Rotation of crops moſt approved of is turnips, barley, clover, and wheat. Fallow practiſed, but not on a large ſcale, unleſs in caſe of turnips. A great deal of bone-duſt uſed, 40 buſhels to the acre, at 18d. per buſhel; but this manure has been uſed to the extent of 80 buſhels per acre, with advantage. Ploughs wrought by two horſes a-breaſt. Large carts and waggons not approved of, and carts of a ſmaller conſtruction thought of more utility to the huſbandman. Wages for labourers are 10s. per week, and a free houſe. Mowing corn from 6s. to 10s. per acre, graſs 3s. No want of hands for harveſt work. Paring and burning approved of on old graſs land; expence 21s. per acre. Country not ſufficiently wooded; a great deal more wanted. The Duke of Norfolk has about 1500 acres of wood in this pariſh; cuts once in 24 years, and leaves a number of trees of different ages each cutting.

From Mr Odey, at Darnhill, near Sheffield, we learned that no regular rotation of cropping was practiſed, and that little land was ſummer fallowed. He farther informed us, that tithes were a great obſtacle to improvements. When he entered to the farm he occupies, four loads of wheat were only produced upon the acre, but owing to the improvements made by him, the produce is augmented to twelve loads; and he conſidered it as a great hardſhip, that the tenth of this aditional produce ſhould be carried off by a man who had born no part of the expence.

Leaving Sheffield, we came to Rotherham, which is a place famous for iron works. Examined ſeveral farms in the neighbouahood, which are generally in good order, particularly that of Mr Taylor at Canklaw Mills. This farm is held upon a leaſe of 21 years from the Duke of Norfolk, and appears under excellent management.

Mr Taylor deals largely in the turnip and graſs huſbandry. His land intended for turnips next ſeaſon had, when we were there, (November 9th) got three ploughings, and appeared almost as clean as many summer fallows. His inclosures are in capital order, all the hedges being neatly dressed, and completely fencible. Keeps a great many sheep, which are of the Dishley breed; and his pastures are of fine quality, being as close at the bottom as if 10 years old, although but newly sown down.

At Aldwark near Rotherham, we received the following information from Mr Wigfull:

The soil about two or three miles round this place, is in general a rich hazle loam, and the climate is warm and dry. The principal proprietors are the Duke of Norfolk, Duke of Leeds, Earl Fitzwilliam, Earl of Strafford, Mr Foljambe, and the Messrs Walkers. But there are also a great number of small proprietors. Farms small in size, being mostly from 20 to 70 acres, and kept nearly in equal proportions of pasture and tillage. The grasses cultivated, are chiefly white clover and hay-seeds. Red clover is sown by itself, and reserved for seed. Not many cattle or sheep bred in the neighbourhood, but a good number of horses since they advanced in price. All kinds of grain are cultivated here; and the general rotation is fallow or turnips, barley, clover, and wheat. The manures used, are stable dung, rape dust, bone dust, horn shavings, &c. Land mostly inclosed, which Mr Wigfull thinks has increased the value of land one-fourth. The wages here are high; ploughmen 10s. per week, besides drink. Labourers 1s. 6d. and 2s. per day. Farm houses and offices are very improperly situated. They ought to be placed, if possible, in the middle of the farm, and not in a corner as at present. The public roads are generally good, but a number of the by-ones are in miserable order. Manufactures of iron and steel, are carried on in the neighbourhood to great extent, which are found to produce good effects upon agriculture, by increasing the riches of the country, and consequently affording a ready market for every article the farmer raises The people have a great turn for improvements, but their genius is cramped for want of leases, and by injurious restrictions laid on them by the proprietors. Tithes are generally drawn here in kind, both small and great. Mr Wigfull suggested that it would be a great improvement in other places of the country, to introduce the sowing of winter tares, which are excellent spring food for horses when their keeping is very dear; and was likewise of opinion, it would be a great improvement in his own neighbourhood, if the corn was cut lower, which would not only take the crop up much cleaner, but also be the means of accumulating a large additional quantity of manure.

The people in this neighbourhood have a great turn for improvements, but are prevented by the following obstacles: want of leases; restrictions in the mode of management, which hinders the farmer from exerting his abilities, and introducing new practices; and tithes, when taken in kind. Mr Hall, at Ickles, informed us, that the tithe of wheat was sometimes commuted for fifteen shillings per acre, when the landlord's rent was only twelve shillings. Mr Hall has a rape mill, and manufactures a great deal of oil, which is generally sold to Lancashire. Purchases rape seed in the East Riding, and Norfolk; present price L. 3 per quarter, and five quarters often raised upon an acre.

Having a letter from Sir John Sinclair to Earl Fitzwilliam, we proceeded to Wentworth-house, but unlukily his Lordship was in Northamptonshire. Delivered the letter to Mr Bouns, his chief steward, who paid us every attention, and from whom we received full information relative to the management of his Lordship's large and valuable estate. Mr Bouns was at the trouble of bringing some of the principal farmers in the neighbourhood to us, from whom we received full and accurate answers to the different queries we had circulated. The following is the substance of the intelligence we received:

Soil variable; both clay upon a wet bottom, and a hazle loam; farms small, not many above L. 100 rent, and chiefly kept in tillage, not above a fourth part being in pasture; grasses cultivated are natural hay-seeds, white clover and trefoil; little red clover sown; both sheep and cattle fed upon the pastures. The cattle are generally of the Craven breed; sheep partly of the polled sort, and a good many from the moors above Peniston. General rotation of crops is turnips, barley, clover, and wheat. Where the land is strong, it is clean summer fallowed, and sown with wheat at Michaelmas; of all the manures that are used, bone dust is found to have the most effect; 60 bushels applied to the acre, and often bought so high as 20d. per bushel. Ploughs and carts are of the common sort; the carts are 7½ feet in length, 3 feet 2 inches in breadth, and 2 feet 2 inches deep, and will hold 1 chalder, or 32 bushels, generally drawn by 3 horses in a line. Few oxen wrought; Lord Fitzwilliam uses some, but the farmers use horses, from their being most expeditious. Land mostly inclosed, the advantages of which are great, being estimated equal to L. 25 per cent.; the inclosures are small, being regulated by the size of the farms; few townships but what have common fields, and these ought to be divided. Not much waste land, but what is of this kind is highly improveable. Wages very high; ploughmen L. 14 a year, besides victuals, drink, and washing; labourers 2s. per day in summer, and 16d. in winter Drains of of various ſizes, and filled with ſtones, but the extent depends greatly upon the goodneſs of the farmer. A good farmer always drains where neceſſary, a bad one neglects it in all ſituations. Paring and burning practiſed upon ſtrong ruſh land, but thought bad huſbandry upon light ſoils. A good deal of wood in the country; but from being too early cut, woods are turning weaker and weaker; cut one in 21 years, a part being left each cutting; ſome trees left to the age of 60 years, a few particular ones longer, moſtly uſed in the collieries. Proviſions at preſent high; beef and mutton 4½d. per lb.; wheat 6s. 6d. per buſhel; barley 5s.; oats 3s.; and beans 6s. Farm-houſes and offices, in general, properly conſtructed for the ſize of the farms and ſtock kept. Leaſes ſeldom granted. No agricultural ſocieties; but the people have a great turn for improvement, the principal obſtacle to which is paying tithes in kind. There are few eſtates in the neighbourhood exempt from paying both ſmall and great tithes, but they are more uſually compounded for than drawn in kind. The greateſt benefits that have been produced from incloſing open fields and waſte land, are in thoſe places where the great and ſmall tithes have been commuted for, either in land or money.

Wentworth Houſe is ſituated between Rotherham and Barnſley, and is one of the largeſt and moſt magnificent houſes in the kingdom. It is unneceſſary here to give any deſcription of it, as Mr Young, in his Northern Tour, has already done this in a very judicious manner. It is ſurrounded by a park, which we were told conſiſted of 1,500 acres, carrying graſs of the moſt exquiſite quality, and upon which large droves of cattle, ſheep, and deer are fed.

Returning back by Rotherham, we proceeded for Parkhill, the ſeat of Michael Angelo Taylor, Eſq. M. P. We were received by Mr Taylor with the greateſt kindneſs:

walked over a number of the adjoining fields with him, and received much valuable information, respecting the husbandry of the neighbourhood.

The soil here is thin, rather wet, and upon lime-stone: Few turnips are cultivated, and they are all sown broad-cast. Mr Taylor showed us a mill for breaking bones, which are in great repute in this neighbourhood, and found to answer better upon lime-stone land than any other manure. Sixty bushels are applied to the acre. Has very little effect the first year, but afterwards operates for a considerable time—we think 10 or 12 years. Prime cost at the mill 18d. per bushel, and the demand greater than can be supplied. Road from Rotherham, till we came near Parkhill, very bad, and all cut into deep tracts; a considerable part of it was almost impassable. Saw some common fields of good natural quality, near a place we think called Maltby, which were under very bad management.

Substance of information received from Mr Arch. Taylor, farmer at Letwell, near Parkhill:

The soil is a thin lime-stone, and the climate moderate. Farms in general too small, which Mr Taylor thinks is the cause they are occupied by a number of poor, bad farmers, as they are not worth the notice of a man of any property. Two thirds of the land is kept in pasture, which is sown with common hay-seeds, white clover, and trefoil, and fed with the Leicester breed of sheep, and long horned cattle. Mr Taylor does not think the long horned kind good for milk, but considers them to answer best upon his thin, wet ground. Mr Taylor's mode of farming is to plough six years, and graze five years. When he breaks up his swarth, applies 2 chalders, or 80 bushels of lime to the acre, and sows turnips for the first crop; 2d, barley; 3d, clover, or pease and beans; 4th, wheat; 5th, clean summer fallow; 6th, wheat with grafs feeds. The firft year of the grafs it is paftured with fheep, and manured in the followed winter; next year cut for hay, from which a good crop of feeds is got; 3d, 4th, and 5th years, it is paftured with fheep. Mr Taylor faid it was not ufual to grant leafes, but thinks a farmer has no encouragement to improve, wanting them. Lands in this neighbourhood fubject both to great and fmall tithes, which, Mr Taylor fays, damps every fpirit of improvement. Mr Taylor ufes a great deal of bone duft, 50 bufhels of which, mixed with fome fhort manure, is fufficient for an acre, although it is an expenfive dreffing, yet as it is very durable, he confiders himfelf well paid for the application. Does not much practife paring and burning, as he confiders it to impoverifh the foil. The land is all inclofed, and has been fo for near one hundred years. Size of inclofures from 5 to 20 acres. Cannot fay whether inclofing has decreafed population or not, being fo long fince it took pace.

From Parkhill to Bautry the road is good. Paffed by Sandbeck, the feat of the Earl of Scarborough, and found the name of the place correfponded with the nature of the foil.

Information at Bautry:

Soil generally of a fandy nature, well adapted for turnips, carrots, and other drilled crops. The land is moftly in tillage, and occupied by fmall farmers and tradefmen. Mr Fifher informed us, he fows red and white clover, and rye-grafs; but that the greateft part of the paftures are fown with hay-feeds, the people having an antipathy to rye-grafs. Rotation of crops here, are turnips, barley, clover, and rye; which anfwers well upon loofe, fandy foil. Manures are dung and bone duft. The fallows are limed with two chalders, or fixty-four bufhels to the acre. Ploughs wrought with two horfes abreaft. Mr Drummond a gentleman farmer here, works oxen. Saw one drawing his water-cart, and working quite calm and docile. The carriages generally ufed, are upon fix-inch wheels, and drawn with three or four horfes. Lands all inclofed, which fets for double rent; but the inclofures by far two fmall. The land here does not ftand much in need of draining, but where it is wanted, the drains are filled with brick. Paring and burning practifed upon new taken-in land. Expence, when done by the plough, 5s. per acre, 13s. when done by the hand, and 2s. for fpreading. Few leafes granted. Mr Fifher informed us he took a farm, and, upon the faith of its not being raifed, made confiderable improvements; but as foon as thefe improvements were difcerned, the rent was raifed immediately; therefore Mr Fifher thinks the want of leafes muft always be a bar in the way of improvements. The tithes are commuted at about 8s. per acre. The great tithes belong to the Duke of Norfolk, and the fmall ones to the clergy. There was lately a fociety at Bautry for improving Agriculture, which did much good, but it has been given up for two years paft.

From Bautry to Doncafter, the land is of a light, fandy nature, upon a wet fpringy bottom. A great part of it has been lately inclofed, but the fences in general are not thriving. Turnips very bad, and little care taken to have the land laid dry, as we obferved much water ftanding on the fields.

Doncafter is a neat, clean town, and there is a deal of fine land in the neighbourhood of it.

Information received at Doncafter from Mr Parkinfon, and Mr Fofter:

There is a great variety of foils in this neighbourhood. A good deal of a fandy nature. Part of it a white clay; and others black earth, or a fine, fharp, light loam. The climate is mild and dry, and both feed-time and harvest are early. The farms are generally small, and mostly kept in tillage. The pastures have usually been sown with natural hay-feeds, but artificial grasses fast coming into practice. Few horses or cattle er bred, and the improvement of sheep but just beginning to be attended to. The rotation of crops upon the light land, is turnips, barley, clover, and wheat; and often a crop of oats taken after the wheat, because there are no leases. Upon the clay land, a clean summer fallow, barley, clover, and wheat; and often wheat taken as the first crop in place of barley. Manures used, are stable dung, lime street dung, bone dust, rape dust, and pigeon dung—about 40 bushels of the last laid upon an acre. Lime applied to the fallow, from 60 to 100 bushels per acre—costs 3d. per bushel. No oxen are used; but this supposed to be owing to the smallness of the farms. Land mostly inclosed, which has produced great advantages. Inclosures from two to thirty acres, but chiefly small. There is a very large common field near Doncaster, of the finest land in England, which is at present let at 31s. 6d. per acre, that Mr Foster thinks would be worth L. 3 : 10s. if divided and inclosed. More than twenty freeholders concerned about it. Their common rotation is, fallow, barley, wheat, and rye, and grass-feeds are sown at different times with all the grains. Another common field is managed differently; the rotation is greatly superior, being turnips, barley, clover, and wheat—the turnips all broadcast, and the most part of them this season are very bad. Upon a third common field, another rotation is adopted, viz. fallow, one half of which is sown with wheat, and the other with barley; then beans and clover; lastly, wheat. And there is a meadow field, which, after being cut for hay, is pastured in common, from the 10th Sepember to the 25th March—above 1200 acres are under the above mode of management. The proprietors are Sir Geo. Cooke, who possesses about one half; Mr Wrightson, who has one-eighth; and a number of small freeholders. Very few leases are granted, which both Mr Parkinson and Mr Foster think detrimental even to the interest of the proprietor himself, as land in that case would set higher. No manufactures here, except one for coarse sacking; but where they do prevail, they are thought to have good effects in encouraging agriculture. Great improvements may be made upon the stock and land in this neighbourhood. Mr Parkinson is of opinion, the horse for the team might be improved by the Derbyshire breed; and that the cattle might be improved, by crossing the Durham cows with the best of the Craven bulls.

With regard to sheep—The Bakewell sort esteemed the best for all the sandy and limestone pastures, and a cross of the large Tees ewes with the Bakewell ram for the strong clay soils. Mr Parkinson thinks the grass land is not sown down properly, being hitherto sown with nasty rubbish called hay-seeds; whereas he is of opinion, it should be done with white clover, trefoil, and rye-grass; and where intended for cutting, with red clover and a small quantity of rye-grass. Thinks also that turnips should be drilled, by which method the land is kept much cleaner, and hoed at far less expence than when broadcast.

Waited upon Mr Childers, at Cantley Lodge, and examined his improvements. The farm in Mr Childers' own possession, which is tithe-free, consists of 520 acres, and by fallowing with turnips, and laying down with plenty of grass seeds, he has made uncommon and substantial improvements. Mr Childers brings manure from

Doncaster, and uses great quantities of lime. He has also a marley clay in his own lands, which he applies to the dry, gravelly, and sandy soils, at the rate of 80 and 100 cart loads to the acre, which produces good effects.

From Doncaster eastward to Thorn, the land is capable of greater improvement than any we have seen in Yorkshire. There is a great deal of common field, superior in quality to most land, and there is also large tracts of waste. At Hatfield there are very large common fields, the rotation upon which is turnips, barley, clover, wheat, and barley; and one of the fields not ploughed, but kept in meadow grass. We examined the turnip field, which consisted, as we were told, of 150 acres, and although of a soil exceedingly proper for that root, they were a crop not worth 20s. per acre. We heard afterwards they were only valued at 15s. The turnips were quite small—few bigger than an egg, and the ground in the most wretched and dirty condition. It appeared to us they had not been hoed at all, or at least very imperfectly, a large proportion was covered with weeds; and worse culture cannot be figured.

If the cultivation was bad, the manner of consuming them was still worse. The whole 150 acres were eating at once, and the flock appeared to be cattle and sheep of all ages and descriptions; such management needs no comment, it speaks for itself.

Betwixt Hatfield and Thorn, there are great quantities of waste land, and much under water. Upon the whole, the land we have seen this day stands in the greatest need of improvement, which cannot be done without a previous division. The common fields to the eastward of Doncaster are abominably crooked and unequal. Some parts of the ridges being twice the breadth of another, and one solitary ridge of wheat often standing by itself—more wretched husbandry could not have prevailed a century ago.

Left Thorn and proceeded northward to Snaith. The greatest part of the land, till we came within two miles of that place, is exceeding wet, and large tracts little better than in a state of nature. The land, though wet and marshy, is generally rich strong soil. Ridges much straighter plowed than is generally the case over the West Riding; but kept by far too narrow and flat. As we approached Snaith the soil turned as fine as could be wished. Great quantities of turnips, and those of good quality.

Snaith is a small market town situated upon the river Aire, not far from its conjunction with the Don. The land round the place is of exceeding rich quality, and but moderately rented. We examined a farm occupied by Mr Latham, and found it well cultivated. Mr Latham, upon his light lands, practises a rotation that has already been often mentioned, viz. turnips, barley, clover, and wheat; but he follows out this rotation in a manner superior to most persons. His turnip crop this year, when so many other people's have failed, is good, and are set to a jobber from Leeds, at L. 6 per acre, to be eat upon the ground. His turnips, although not drilled, are all in rows, about sixteen inches wide, which enables him to hoe them with accuracy. His method to do this, is to give the last furrow very broad, which takes all the seed when harrowed into the furrow, and so gives the field an appearence of regularity. Mr Latham said this plan was fallen on by accident, which indeed is often the parent of many improvements;—when ploughing one of his fields some years ago, he ordered his servants to finish it that night. There being a feast in the neighbourhood, the ploughmen were anxious to be early at it, and so gave a furrow much broader than usual. When the young plants came up, Mr Latham was surprised to see them in regular lines, and inquired into the cause of it; which pleased him so well, that he has since continued the practice.

Mr Latham sows rape upon his wheat stubbles, that are next year to be turnips. His method is to plough the field as soon as the wheat is carried off, and sow the rape immediately, which is generally got down by the middle of September, and affords him feeding for his sheep in spring equal in value to 20s. per acre.

A part of Mr Latham's farm is what is called warp-land, or land enriched with the sediment left by the river Aire, when its banks are overflown. Upon such fields he does not venture to sow wheat, as it stands in danger of being perished; but from the richness of the soil great crops of spring corn are raised.

From Snaith to Ferrybridge there are a number of common fields, which were under no better management than those we have formerly described. We saw a large common field of turnips to the eastward of Kellington, which were middling good, but very imperfectly cleaned. At least 40 acres were flocked off at once, and cows, bullocks, young cattle, and sheep were feeding indiscriminately. Saw also upon this road some fields of rape intended for feed, which looked well.

Waited upon Richard Slater Milnes, Esq; M. P. for York, at his house at Fryston. From his information, and that of others, the following account of the husbandry in the neighbourhood of Ferrybridge is given:

The soil is composed of lime-stone, clay, sand, &c. in the vallies; and rich pasture, and meadow lands near the rivers. The land is chiefly possessed by large proprietors; such as Lord Mexborough, Mr Milnes, Mr

Crow, &c. Farms contain from 50 to 300 acres, and mostly kept in tillage. Large quantities of red clover and sainfoin are sown for cutting, which answer well; and white clover, trefoil, and hay seeds are sown for pasture. Some lucern is sown, but the quantity inconsiderable; many horses are kept on account of the coallieries, lime works, drawing vessels along the river Aire, and other purposes besides that of husbandry; which consume the red clover and sanfoin. The pasture inclosures are generally stocked with sheep; and the lands near the water side are eaten by milch cows.

Rotation of crops upon the clay land is, wheat or barley upon the fallow, and afterwards oats, or beans. No more than two crops are taken to a fallow, unless the land is of superior quality. Turnips are sown upon the light land, and followed with barley, clover, and wheat. The manures used are, stable dung, pigeon dung, and sometimes bone dust. A great quantity of lime stone is burned at Knottingley and Brotherton, which is laid on, from two or three chalders per acre. Ploughs are of the usual kind kept in the country, and generally drawn by two horses a-breast. Carts with wheels of 3, 6, and 9 inches broad, and drawn by three, and by four horses in a line, are used. Very few oxen are wrought, and those only by gentlemen. Land mostly inclosed, but the inclosures thought too small. Inclosing is reckoned to produce the following advantages: It enables the possessor to cultivate the land in a superior style, which, in its open state, it was out of his power to do. From such cultivation a greater produce is obtained; and on the light soils the turnip, clover, and seed husbandry cannot otherwise be practised to advantage. Provisions are cheaper here than in the manufacturing part of the country, at least one halfpenny per pound. Roads, both turnpike and bye-ones, are in good condition.

Sometimes the affeffment for fupporting them is 1 S.l. per pound upon the valued rent. Much improvement has been made by draining, and great attention paid to it. The farm houfes and offices are in general very inconveniently fituated, moft of them are in villages, which of courfe renders a number of them at a great diftance from the land. Some leafes are granted; but it is not the general practice to give them for more than *one year*. No modes of hufbandry prevail here that would be of advantage to other places, except fowing fainfoin, which anfwers well upon all chalky, or limeftone land. Some bills have paffed for dividing common lands, which have produced great advantages. Plentiful crops have been raifed at little expence, and an opportunity given of laying down the old going land into grafs; alfo an exemption from tithes is procured by them.

We proceeded to Selby. This is a populous market town, fituated upon the river Oufe, and was the birthplace of King Henry I. on which account his father William the Conqueror built an Abbey here. From John Fofter, Efq; we received the following important intelligence relative to the agriculture of this part of the country:

The foil is various, part of it fandy, and part a hazle clay. The climate is moderate; the proprietors are Lord Petre, the Archbifhop of York, and a great many copyholders. Farms are fmall, and kept in equal proportions of pafture and tillage. All forts of graffes are cultivated; which are ufed both in breeding and feeding. Sheep are generally of the Northumberland kind, and the cattle of the fhort horned or Holdernefs breed. Great numbers of horfes are bred. The rent of the land is from 5s. to 50s. per acre.—Rotation of crops: when land is broke up from grafs, flax is generally the firft crop, then rape, afterwards wheat, and a fallow;

but no fixed rotation is kept. Ploughs of the common kind, drawn by two horses are used; but a number of oxen are wrought in the waggons. There are no common fields in this parish, but many in the neighbourhood. The difference of value betwixt open and inclosed lands, is estimated at one-third, or 33 per cent. Here is a considerable deal of waste ground, which produces little or nothing at present, but is capable of great improvement. Strict attention is necessary in keeping the ditches clean, and letting the water off the fields, which are greatly hurt by rain water stagnating upon them; but as there are no spouts, little other draining is required. Provisions are plenty and moderate; roads tolerable, great improvements have lately been made upon them. Farm houses and offices are well enough constructed, but very improperly situated, as they are mostly in villages. A number of landlords do not grant leases, which is destructive to good farming.

We proceeded for Tadcaster. Great part of the country is upon a lime-stone, and lies very well: but the ridges in general are too flat, and no attention paid to letting off the water. We saw several common fields. After passing Sherborn (at which place great quantities of the Winesouer plumb grows), the country appeared very thinly inhabited; few or no houses being to be seen, till we arrived in the immediate neighbourhood of Tadcaster.

At Tadcaster we were recommended to a Mr Potter, as one of the best farmers in that place; and we found that his practice was accurate and correct, in the highest degree. We received the following information from him:

The soil is a dry lime-stone; the climate kindly and moderate. The proprietors mostly have large estates; but the farms are small, few extending to 300 acres.

The greateſt part of the land is in tillage, not above one-third being in paſture. The graſſes ſown, are red and white clover, trefoil and fainſoin. Rye-graſs is out of repute, and hay-ſeeds faſt following. Sheep are kept upon the paſture land, and cattle fed upon turnips. No land is floated or watered. General rotation of crops is, turnips, barley, clover, and wheat; often a crop of oats taken after the wheat. The manures uſed, are dung, made upon the farm, and gathered at Tadcaſter; ſome lime brought by water from Hull, and horn ſhavings from York. The ploughs are of the Dutch kind, and drawn by two horſes a-breaſt. No oxen uſed, but thoſe kept by Lord Hawke. The ſowing of wheat commences about the end of September, and continues all the month of October. Spring crops are ſown as early as poſſible. The harveſt is early. Here are ſome common fields; and Mr Potter ſuppoſes, the difference of value betwixt open and incloſed land to be one-fourth. Incloſures are ſmall, few exceeding ten acres. There is a good deal of waſte land, ſome of which is under diviſion, and capable of great improvement. The wages of a labouring man is 9s. per week; ploughmen get L. 10 per year, beſides victuals and waſhing; the head man gets equal to L. 30 per year. Hours of labour are ten in ſummer, and ſeven in winter. Paring and burning are very ſeldom practiſed. A great quantity of the Wineſoner plumb is produced in this neighbourhood. Mr Potter thinks it would be highly beneficial to the public intereſt, that all land was ſet under leaſe; and further thinks, there is no neceſſity for impoſing reſtrictions on the good farmer, as he will manage much better wanting them; and as for the bad farmer, he cannot be mended by them. The people here have a great turn for improving their lands; but have no opportunity of doing this to purpoſe, from the want of leases. He thinks the small size of the farms serves to retard good management.

Waited upon Mr Beck, steward to Lord Hawke, upon his estate of Scarthingwell and Towton.—His Lordship has taken about 1600 acres into his own hands; and is very properly putting it into good order, by fallowing, manuring, and, laying parts of it down with grass seeds, with a view to let it in proper sized farms to substantial tenants. Besides the manure raised on this farm, his Lordship has expended yearly above three hundred pounds in purchasing manure, principally dung, from the towns and villages in his neighbourhood, and by water from Hull, York, &c.

The soil upon Lord Hawke's estate is of many different kinds: it is good loam in general: there is also clay upon limestone; strong clay upon a blue till; hazle earth upon sand; and about 50 acres of moss, or peat earth. About a fourth part is kept in pasture, though less pasture in general is kept. Lord Hawke cultivates sanfoin, red clover, and trefoil, with white clover, and hay-seeds. He bred 350 sheep last year, and has this year increased his breeding ewes to 440: they are of the Oxford and Gloucestershire polled breeds; they have a cross also of the Bakewell and Fowler breeds; and the wethers are fed off when shearing. at 38s. each. He folds his ewes always from May-day to Michaelmas. He feeds also a few Scotch and Irish cattle. The general rotation of crops is turnips, barley, clover, and wheat. His plan now adopted, is to sow half his clover land with twelve pounds of red clover per acre; to mow it once, and then feed it. The other half is sown with 6lb. of white clover, 3lb. of rib-grass, and 6lb. of trefoil per acre, and fed, but not mown. By this rotation of crops, red clover is sown but once in eight years on the same land. His plan is to lay down one hundred and fifty acres with fainfoin, the feed of which he fows with his barley; and has fometimes fowen it on a clean fallow, when the ground laid down with fainfoin would have been broken up for wheat had it been fown with clover, he breaks up an old worn-out pasture ground, and fows it in the fpring following with oats; after which it is fallowed, and falls regularly into courfe, inftead of the ground fowen down with fainfoin. The manures ufed, are rape duft, pigeon, farm-yard, and bought dung, foot, rape, and bone-duft. Lord Hawke ploughs with two oxen abreaft, without a driver, and fometimes with horfes, but depends principally, and almoft entirely on oxen, for his ploughing and harrowing. His land in hand is all inclofed; inclofures vary from 8 to 30 acres. There are fome paftures from 5 to 6. We think fmall arable inclofures hurtful in a corn country; and Lord Hawke is altering the fize of his fields, from 15 to 20 acres. Mr Beck is of opinion that inclofing is very beneficial, and never can decreafe population. Lord Hawke had land in a common field, for which he got only 5s. 9d. per acre, and can let the fame land, when it is now divided and inclofed, at 20s. Wages are high; houfe fervants coft, in board and wages L. 30 per annum. Daining is much required here; but for want of a law to oblige neighbours to clean out their contiguous ditches, it cannot be done to advantage; although Lord Hawke is attempting it, and has induced many to drain with him. Paring and burning are practifed on old grafs land, and thought an excellent method of breaking up all coarfe fward. Lord Hawke approves of it on low grounds, but on high ground thinks burning unneceffary, and rather detrimental.

Left Tadcafter and took the road weftward to Harewood. Obferved fome common fields by the way. The land in general is upon a wet bottom; and from the rains, and the little attention paid to clearing out the furrows, is in a very bad fituation.

We delivered a letter to Mr Samuel Poplewell, fteward to Lord Harewood, and received fatisfactory information from him. Harewood is a neat village, and his Lordfhip's refidence is a little diftance from it. He grants no leafes, but is efteemed a kind landlord.

The foil is generally clay, upon a bottom retentive of moifture; the climate fhowery and wet. Land is chiefly poffeffed by large proprietors, and occupied by tenants paying from L. 20 to L. 200 yearly rent. It is employed both in pafture and in tillage, in proportions nearly equal. The paftures are moftly eaten by fheep, which are purchafed from Northumberland; their fleece fells from 3s. 6d. to 4s. Many Scotch and Irifh cattle are fed upon the fides of the river Wharfe. Upon the tillage land two crops are generally taken to a fallow, and turnips fown upon all the fallows proper for them. Mr Poplewell drills his turnips, and has never miffed a crop fince he practifed that method. The manures ufed are, home-made dung, rape duft, rape coombs, and dung and foot from Leeds. Little lime is ufed, excepting on new broken up land. Ploughs are generally drawn by three horfes in a line. No oxen are ufed for work, excepting a few by Lord Harewood. Some rape is fown, which is often eaten by fheep, but fometimes ftands for feed. Here are no common fields, but there are fome in the neighbourhood, which Mr Poplewell thought fhould be divided. He eftimates the difference betwixt open and inclofed land, to be at leaft 25 per cent. He alfo is of opinion, that it would be of great fervice to agriculture, if all lands were fet under leafe; and that if thefe were granted, there would be no neceffity for reftrictions, unlefs during the concluding years. A bill paffed about three years ago, to divide a common in this neighbourhood, which has produced beneficial confequences; and Mr Poplewell is of opinion, moft part of the wafte land in the Riding might be improved, by planting Scotch firs upon it.

We arrived at Wetherby, which is a great thoroughfare on the London road. Here we received the following information:

The foils in this neighbourhood are lime-ftone and ftrong clay. There are a few fmall freeholders, but the land almoft wholly belongs to the Duke of Devonfhire. Farms are generally fmall, the moft part not exceeding L. 30 per annum. Rent is about 20s. per acre, and the public burdens. Rotation of crops upon the lime ftone is, turnips, barley, clover, and wheat; on the clay, fallow, wheat, and beans. The manures ufed, are great quantities of rape duft, price 2s. 4d. per bufhel; horn fhavings from York, foot, and all the dung that can be collected at home. Lime is applied to the fallow, 100 bufhels to the acre; it cofts 9s. 6d. per chalder of 32 bufhels. Ploughs are of the common kind, and drawn by two horfes upon the lime-ftone, and by three and four upon the clay land. No oxen are ufed. Harveft is early; begins generally about the firft of Auguft, and is all finifhed by the middle of September; the land is all inclofed; the fize of inclofures from 3 to 12 acres. Wages are high; ploughmen, that are mafters of their work, get fifteen guineas per annum, befides victuals; and labourers never lefs than 18d. per day, and more in harveft; no fcarcity of hands to reap the crop, excepting in the year 1792. The corn is moftly cut with the fickle; wheat is done for 7s. per acre. Provifions are plenty, but high priced. Farm-houfes and offices are improperly fituated, as they are all placed at the corner of the lordfhip.

The Duke of Devonfhire formerly granted leafes, but now intends to act otherwife; which we were told would be a great bar to improvements. The covenants that formerly fubfifted were, to keep two-thirds in grafs, &c. Tithes are generally commuted here, and 7s. per acre paid in their place.

Ripon is of great antiquity; being, it is faid, incorporated by King Alfred; and is pleafant and well built. The river Eure was made navigable about twenty years ago, and a number of veffels are employed thereon, to the great convenience and benefit of this place and neighbourhood.

The foil near Ripon is partly of a fandy nature, and partly ftrong clay upon a limeftone; the climate healthy, and moderate. Eftates are generally large, and farms of various fizes, from L. 20 to L. 300 yearly rent. The lands are moftly in grafs and meadows, little more than the fourth part being kept in tillage. Artificial graffes are juft beginning to be introduced into the hufbandry of this neighbourhood. A few cattle of the fhort horned kind are bred, and a good many long woolled fheep, which when fatted at two years of age, will weigh 25 lb. per quarter. The rotation of crops is, turnips, barley, clover, hay-feeds, and wheat, upon the light and fandy foils; and on the ftrong foils, fallow, wheat, and beans. Lime and common dung, with a little rape duft, are the only manures ufed. A large heavy plough, drawn by 4 and 6 horfes, yoked in pairs, is employed upon the ftrong lands. Upon the light foils, a fmaller plough drawn by 2 horfes is ufed. The country is moftly inclofed. Inclofures are from 5 to 40 acres. Mr Peacock thinks, land when inclofed is of double value, to that of fimilar quality, when lying in common field,

There are some thousand acres of waste or common in the neighbourhood; most of which is capable of great improvement. Wages for labourers are at 2s. per day in summer, and from 1s. to 1s. 4d. in winter. Little of the country requires draining; but where this improvement is necessary, it is well attended to, plenty of materials for this purpose being at hand. The average price of butcher meat is $3\frac{1}{2}$d. per pound.

Farm-houses and offices lately erected, are in general good, and conveniently situated; but those that have stood long are not so. Mr Peacock thinks, that the principal obstacles to improvements are, the want of leases of a proper duration, and the restrictions from ploughing up the old grass fields, which effectually prevents any new systems of husbandry from being introduced.

No II.

ACCOUNT OF THE VALE OF SKIPTON,

IN A LETTER FROM A GENTLEMAN IN THAT NEIGHBOURHOOD.

GENTLEMEN, Nov. 9, 1793.

It would have given me particular pleafure, as a fincere well-wifher to your undertaking, if I could have acquitted myfelf more fuccefsfully in the inquiries you honoured me with; but I fhall give you the beft information in my power.

With regard to the ancient ftate of this vale, I do not find, upon inquiry, that there has been any material alteration or improvement for the laft century or more: in fome parts of Craven, though not near this town, I underftand that, even within the laft forty years, there was a confiderable portion of land in tillage; the ploughing was then performed by four or fix oxen, and one or two horfes; and I am informed that mode of hufbandry anfwered very well. Craven was then famous for a breed of long horned cattle, particularly oxen; but fince the introduction of Scotch cattle and grazing into the country, the long horned breed, and of courfe the tillage has been neglected. One caufe of this is the eafy expence that attends this mode of hufbandry; with one fervant, and two horfes, a farmer can very conveniently manage feven or eight hundred acres of land; indeed, moft of the grazing farms in this vale are very large, often three or four are united under one occupier.

The Earl of Thanet is the principal proprietor of land in Skipton; and, I am told, is not willing that his fine land should be ploughed; but it would certainly be a great advantage to the neighbourhood, if a proper mixture of grazing and tillage could be introduced; for though the country is not, or ever will be populous, while the present mode of husbandry and monopolizing farms prevail, yet corn is generally higher in Craven than in most parts of the kingdom, because so very little is produced. If you suggest to them, that the uplands may be kept in tillage; the reply is, that they are so much exposed to mists, and the situation so cold, that corn, particularly wheat, cannot feed or ripen. This may be in part just; but the stronger reason with them seems to be, that the uplands are very useful to them upon their present plan, to prepare the lean cattle for the better pastures; which some say, would be too rich for them in that state; nor would their improvement, at first, be equal to such keeping.

The extent of the Vale of Skipton cannot be accurately ascertained; indeed, a very small part bears that name, being generally included in the vale of the river Aire which extends from Leeds, in a north-west direction, to the source of the river, about thirty-five miles, is upon the average about a mile broad, in some places more, yet not so much (I think) as to add a quarter to the average. Grazing is the general mode of occupation in this vale, except in the neighbourhood of the manufacturing towns, where conveuience will command a higher rent than the grazier can afford to pay. Six pounds per statute acre, and sometimes more, will be given for land in such situations: grazing will not answer to half that price.

It is worthy of notice, as it appears to me of great service to the land, as well as very profitable to the occupier, that moſt of the principal graziers take all their ſtock out of ſome of their beſt paſtures in the beginning of July, and put nothing in them till about Michaelmas, when they are equal or ſuperior to the beſt fog; indeed they call this, fogging their paſtures. The favourite grazing ſtock here, are the black Scotch cattle, ſome ſheep; but on the lowland very few, and on the uplands and moors not very numerous:—it is much to be wiſhed that the propagation of this uſeful and profitable animal was more attended to.

Price of labour. A man ſervant about ten guineas per year, with board and waſhing in his maſter's houſe; a woman about five guineas, with the ſame; day labourers in huſbandry about 2s. or 2s. 6d. per day, finding their own victuals: about ten years ago, 1s. or 1s. 2d. was the common price; the advance owing to the introduction of the cotton manufactory into a country ſo little populous. They work from ſix to ſix in ſummer, and from eight to dark in winter.

Price of proviſions for the laſt year: beef, mutton, veal, and pork, about 4½d. per pound, of 16 ounces; butter about 1s. or 1s. 1d. per pound, of 22 ounces: wheat about 8s. per Wincheſter buſhel: oats 28 or 30s. per quarter.

The climate and weather are unfavourable: we have ſometimes very cold eaſt winds in the ſpring for three months, often to the middle or end of May; in autumn we have very often heavy and continued rains from the weſt, owing to our ſituation among ſo many hills; from the ſame cauſe, we have frequent thunder ſtorms in ſummer.

Our roads are very much improved of late; the canal which is carried through this valley, ſeems to have taught us the poſſibility of making tolerably level roads, even in a mountainous country; ſeveral excellent ones have been made within the laſt five years; the materials chiefly lime-ſtone, broken to about the ſize of an egg.

Tithes are generally collected in kind, and are very reluctantly and ill paid. Since the introduction of grazing into the country, they are reduced in an aſtoniſhing degree; the lands which are moſt profitable to the occupier, are leaſt, or indeed not at all ſo to the clergyman;—he muſt either ſubmit to this, or involve himſelf in a tedious and expenſive law-ſuit, for agiſtment tithe, perhaps againſt an obſtinate and powerful combination of the farmers and land-owners. It is the opinion of the moſt intelligent people here, that the preſent mode of collecting tithes is one principle cauſe of the high price of corn. Large quantities are continued in graſs, which would be ploughed to advantage, if a certain and general commutation for tithes could be eſtabliſhed. I wiſh the above hints may be of any ſervice to your buſineſs; if you think me capable of further information, I ſhall always be happy to contribute my aſſiſtance to ſo laudable an undertaking. I am, &c.

No III.

EXTRACT OF A LETTER FROM WM PAYNE, Esq, of Frickley near Doncaster. *Dated Nov.* 30, 1793.

GENTLEMEN,

I LAST week saw your queries on the state of Agriculture in the West Riding, inserted in the Doncaster paper, and have taken the liberty of answering them, according to your request, in the address that precedes them. And having understood, that the indefatigable President of the Board of Agriculture was desirous of obtaining a *detailed* survey of England, I shall principally confine my replies to your inquiries to the parish in which I reside, Frickley cum Clayton, and the extensive and populous one adjoining it northward, South Kirkby. Yet these answers will, I believe, *generally* apply to the whole tract of country lying between the market-towns of Doncaster, Rotheram, Pontefract, Barnsley, and Thorne; in divers parts of which district I have resided, and practised agriculture, as a freeholder; not having been without the means and inclination of acquiring some intelligence in many departments of its rural economy. As a true friend to the solid prosperity of my country, I am a sincere well wisher to its agriculture, as the only sound basis of its real and permanent interest; and though I do not wish manufacture in general to be depreciated, yet I am convinced, that if a considerable portion of the public industry and capital which for some years past has been applied to the manufacture of foreign materials, had instead thereof, been employed in the cultivation of our extensive wastes, the profits on the *whole* of such employment to the *public* would have been immensely superior. On this view of the subject, the institution of a Board of Agriculture may be important to the national welfare, if the public spirited activity of true patriotism abounds in its members; but if there is not a degree of that liberal principle, sufficient to promote, and obtain some modification of certain impediments to the extension of our agriculture, the attention of the Board to any other means of exciting and encouraging rural industry will ultimately be contracted, by the mere expedients of the day, and the labours of its useful members prove in vain.

The soil of this district is of three kinds, with their varieties, viz.—1st, A dry loamy hazle soil, on a rock of soft gritston; 2d, A wet or clay soil, which abounds most here; 3d, A fine dry loam, on a rock of limestone. I think the climate more favourable and mild than in some other parts of the county, with less rain.

Nearly three fourths of the lands are employed in tillage, the other fourth part is chiefly clay land, meadow, and pasture: but the practice of ploughing old swards, and laying new ones, prevails on all the soils. Red and white clover, trefoil, with common hay seeds, not of the best sort, and sainfoin on the lime-stone soils, are cultivated as grasses. The common rotation of crops, on the drier soils, is: 1st, turnips; 2d, barley; 3d, clover or beans; 4th, wheat: On the wet or clay soils; 1st, fallow; 2d, wheat; 3d, oats; 4th, wheat; sometimes the course is, 1st, fallow; 2d, barley or oats; 3d, clover or beans; 4th, wheat; which is generally esteemed the better course; in a few instances, potatoes and cabbages are cultivated in the lieu of turnips.

Summerfallowing is universally practised on the dry soils

and good ſpring dreſſings on the drier ones, for turnips, &c. Turnips are generally ſown broadcaſt; but the expertneſs of our hoers ſufficiently compenſates for the want of drilling. That excellent mode of cultivation, the hoeing of turnips, has been practiſed in this part of the country upwards of thirty years; being introduced about that time into the townſhip of Wath upon Derne, by that excellent cultivator, William Payne of Newhill Grange, my late honoured father; as it was to the county, by that truly patriotic nobleman, and benefactor to his country, the late Marquis of Rockingham. Yet I am ſorry to obſerve, this moſt beneficial practice is ſtill much neglected in ſome parts of this Riding, particularly in the neighbourhood of Thorne and Hatfield.

The manures uſed here are: 1ſt, farm-yard rotten muck; from eight to twelve 3-horſe cart load of which are applied to the ſtature acre of fallow; 2d, aſhes, about eight loads per acre; 3d, ſoot, chiefly as a top dreſſing for wheat, from twenty or thirty buſhels per ditto; 4th, Bone duſt and horn ſhavings, from three to five quarters per ditto; 5th, dove manure, ditto; 6th, ſoap aſhes, ditto; 7th, rape-duſt, ditto. Lime is generally employed as a manure for the firſt fallow after an old lay, apparently with ſucceſs, at the rate of two or three chaldrons per acre. My own practice for turnips is, one chaldron of lime well mixed with the ſoil, and ſix loads of freſh muck, or three quarters of dove manure per acre, with full ſucceſs; this compound manuring, I think, inres its due operation on the ſoil in moſt caſes better than the ſimple one, and has many other advantages. The ſheepfold is not uſed here, except on turnips, which are generally eaten on the land by ſheep.

The common ſort of both broad and narrow wheeled carts, with three or four horſes, are generally uſed, with a few one-horſe carts; ſcarcely any other plough is ſeen than the common ſingle one. The work is almoſt entirely performed by horſes; very little uſe is made of oxen at preſent; though where they are employed, they are found to anſwer very well, and I have no doubt of their ſuperiority over the heavy draſt horſes in point of *real* utility to the farmer. I have uſed a pair of oxen ſeveral years in harneſs like that of the horſes, working them at the plough and on the road, in every reſpect as we uſe our heavy draſt horſes; and as far as I can judge, they are equal to them for *uſe*, though the pride of the drivers will never allow it. However, in the ſtage of fattening them, we are all agreed, that *their beef* is preferable to the *carrion* of an old horſe. The advantage to the community of working oxen on farms is beyond diſpute, or calculation.

The rate of wages is low, the price of neceſſaries conſidered; and hands for the purpoſes of agriculture, in its preſent imperfect ſtate, are not wanting.

Paring and burning are practiſed generally on the breaking up of old lays, the expence of which is from 16s. to 21s. per acre.

Proper attention is paid to the draining of *arable* lands, but I cannot ſo fully anſwer for it in other reſpects.

Few leaſes are granted, and I rather think few are aſked for; the nature of the covenants between landlord and tenant, has a general reference to law and cuſtom, which ſecure to the landlord quiet entry on due notice, with recovery of damages if any be done to the farm; and to the tenant, on quiting, a fair valuation of his property and labour, in the ground; as fallows, crops, manure, &c. &c. being parts of his *ſtock* in *trade*. It is an article eſſential to a good and ſpirited agriculture, and which cannot be too much inſiſted on, that the farmer be ſcrupulouſly allowed, on quiting his farm, a fully and fairly appraifed valuation of his *flock* in *trade*. It forms a fecurity and bond of entire confidence, equally to landlord and tenant, a fecurity which fets all leafes, parchments, bonds, and feals at defiance; it fecures to the landlord the payment of his juft demands, with a certain improvement of his eftate: and to the tenant an eafy mind, under the application of his ingenuity, induftry, and cafh, to the profpect of increafing his produce, and ameliorating his farm. I wifh this matter was more attended to; I have feen many painful deviations from juftice in this refpect, to the great injury of the *caufe*. An act of the legiflature might probably extend this *real* benefit, and promote the improvement of the lands already inclofed, more than *millions* expended in the way of premium, &c.

There is no other obftacle to improvement but the payment of tithes in kind; an obftacle, the effects of which upon agriculture might be much diminifhed, if not *entirely removed*, if the Members of the Board could unite their labours in fo important a caufe, with a fincere zeal and regard for juftice, and the religion of Chrift. The obftacles to the improvement and inclofure of wafte lands, in many places, amount nearly to a *prohibition*; viz.—1ft, The tithes, the diflike of which, with the freeholders, &c. makes a very difficult commutation, the abfolute condition of their concurence. 2dly, Manorial claims and powers. 3dly, The heavy expence and trouble of obtaining acts of the legiflature. To which may added, the caprice, *partial* intereft, and difinclination to all improvement of fome of the claimants in many cafes. All thefe obftacles might be much leffened by a law, fpecifying and *explaining* the claims, and *limiting* the *powers* of tithe and manorial proprietors, in fuch manner, that *their fimple oppofition* fhould not hang *in terrorem* over the very threfhold of every fuch inclofure; and alfo *facili-*

tating and *encouraging* fuch applications to the legiflature; perhaps a general act of inclofure upon a good plan might be a wife and feafonable meafure to liberate the *active improvers* from the torpid dominion of indolence and ftupidity; however the government can fcarcely do wrong in this matter, except by *fuffering* the *waftes* to remain as they are.

Entirely owing to one or all of the obftacles I have mentioned, very few indeed of inclofing bills have paffed thefe twenty years, in the whole diftrict comprifed between the towns I mentioned above, notwithftanding the value of the lands, and the great fcarcity and fmallnefs of farms; in the few inftances that have occurred, their beneficial confequences to the ftock of public induftry and produce have been confpicuous.

Tithes are drawn in kind here, and generally over this diftrict; yet there are fome inftances of payment in money by annual agreement, &c. If genuine chriftianity, if agricultural profperity; if domeftic peace, and fmiling plenty, be for the public good; then it will be for the public good to have the tithes commuted, and their very name abolifhed for ever.

No IV.

EXTRACT of a LETTER from a FARMER in the neighbourhood of Pontefract.

Dated Dec. 14, 1793.

GENTLEMEN,

THE land betwixt Doncaster and Ferrybridge, is chiefly lime-stone, or gravelly soil. All along the road there are many open fields, which are capable of great improvement, by inclosing, sowing grass-seeds, and pasturing with sheep.

The present tenants are in general poor, and the farms small; poverty causes a kind of stupidity to take possession of them; and I have often spent my time in attempting to convince them of their errors; but though many of them may be convinced, it is not in their power to get out of the old mode, for want of the one thing needful.

The lands I allude to, are chiefly in tillage, the labour of which, and the necessary manure eats the poor tenant up.

Westward of this road, we have useful land, that can feed cattle and breed good sheep. Mr Sayle has done much good in these respects; some of us are following him as fast as we can in the Dishley breed; but he has got the lead, and I wish him success, for he deserves it.

Common hay-seeds are going out of fashion with the

beſt farmers, and clovers and rye graſs daily gain ground. Sainfoin is very uſeful in barren or poor lands, and in good ſeaſons, as great crops of it are produced, as we can have of other graſſes upon our moſt fruitful ſoils: happy it is for the occupiers of ſuch land, it was found out. We have a very indifferent breed of cattle. If gentlemen would ſend good bulls amongſt their tenants, and let them ſerve their cows *gratis*, it would, I think, be the only probable means of attaining ſucceſs in this moſt eſſential point.

There is no land watered here, but many ſituations are well adapted for that purpoſe. I myſelf have 30 or 40 acres, which I have long wiſhed to float, but as I have no leaſe, the expence deters me.

The ſucceſſion of crops we have after fallows, is barley, clover, and wheat. Turnips are taken wherever the land is proper for them; but we have not much of that kind hereabouts. Upon ſtrong lands, we ſow wheat after fallow, then beans, and conclude with wheat or oats. Tares are now coming in faſhion.

Oxen are not much uſed for work here, and never will become general; as they are thought too ſlow by the active farmer.

There are many fields open over the country, which would be far more valuable if incloſed; alſo ſeveral common waſtes, to which the ſame obſervations will apply.

The advantages ariſing from incloſing are obvious, by an increaſe of labour, and an increaſe of food, both of which are for the public good. It produces diſadvantages to none, unleſs it be a few individuals. In the village where I live, and where we have had no incloſure bills, the increaſe of poors rates has been incredible. I am not very old, and can remember the time, when we had only one poor woman upon us at 6d. a week; but for theſe ſome years back, the expence of ſupporting the poor has been from L. 150 to L. 180 a year; and this chiefly paid by tenants not renting above L. 1,600 all together.

Wages are much advanced. I have two labourers, which coſt me not leſs than L. 60 a year: in ſhort the expence of labour is become unſupportable. Draining is uſed often among us; perhaps more might be done if it was not a heavy expence. Paring and burning are alſo uſed, and are without doubt an excellent practice on ſome lands. I have no notion it waſtes the ſoil, which is the chief objection our young agriculturiſts have againſt it. The expence is from 20s. to 28s. per acre.

The modern farm houſes, and offices, are much ſuperior to thoſe formerly built. I would have every farmer reſide in the middle of his farm; and every houſe and home-ſtead built in an uniform and convenient manner.

Leaſes are not univerſal enough for the encouragement of experimental agriculture; and the nature of the covenants is according to the liberal or illiberal diſpoſition of the landlord. One will ſmile upon the arts, and lead rural induſtry by the hand, whilſt another caſts a damp upon the honeſt heart by oppreſſion, and clips the wings of riſing genius.

With regard to improvements, ſome have the will but not the power to make them; others the power, but not the will. Nothing but numerous and repeated examples can influence the ignorant and ſtupid. Thoſe who have the inclination, but not the means, ſhould be aſſiſted by their landlords, and pay poundage for it. Where land is to be watered, this ſhould particularly be the caſe, for it will enable the tenant to pay intereſt with a ſmile.

Cabbages might be grown upon many lands improper for turnips; and if planted with intervals of four feet, as at Diſhley, the ground would be kept clean at little expence. I have found them exceedingly ufeful. No land fhould lie dormant for a year; and if no man plough-ed more than he ought, he would always be enabled to turn his fallow brick to fome ufeful purpofe.

No V.

The following Extracts from Two Letters, written by a Farmer in the West Riding, contain so much natural good sense, expressed in forcible language, upon the Obstacles to Improvement, and the means necessary for rectifying the practice of the Husbandry in that district, that we have given them a place. At same time we beg leave to observe, that this gentleman's sentiments, so far as they go, are nearly similar to those we have formed in consequence of our Survey.

Being desirous to encourage an undertaking which has for its object the improvement of agriculture, and of course the general benefit of the public, I have ventured to communicate my thoughts to you upon some of the most important obstacles to that useful science, which I thought might be more clearly done upon paper than in the short time I had the honour to spend with you. I will begin my observations upon the third of your queries. The lands in this part being chiefly occupied by small farmers they are deprived of making that improvement which a man of property, with 200 or 300 acres of land, can do. A small farmer, not having room to change his land from tillage to seeds, and pasture with sheep, which is the grand improvement of the land in this part, he lets a small portion lie in grass, to keep his milch cows and horses, and the rest perpetually in tilage, excepting now and then a little broad clover. By this method it gets wore out, requires a heavier dressing of manures, more working by the plough and harrows, and becomes so fixed and cemented together (the greater part of our land being of a clayey glutinous nature), that it is deprived of receiving the benefit of the sun and air, which is the principal life of vegetation; whereas by laying down with red clover, and white and yellow clovers alternately, and occasionally a few grass seeds, the soil is kept in a freer state. The fibrous threads of those seeds running among the soil, communicate the warmth of the sun and air in every part, render the soil more malleable, easier to work, and in a better state for the reception of any kind of grain. These advantages it receives from the culture of seeds, exclusive of the *rest, and the manure*, which is scattered upon it by that most provident of all cattle, sheep, as great a portion of which I should recommend to be kept upon every farm as is consistent with this mode of management. They enrich the soil more than any other cattle; and give employment by their fleece, and are the most approved food in their carcass, to our manufacturers. Another obstacle to improvement here is, that a small farm is not worth the attention of a man of ingenuity and property; and this, together with the refusal of leases and arbitrary clauses, prevents men of property from educating their sons in this line of business. Every man therefore that experiences these oppressions, and who can give his son a fortune to stock 150 or 200 acres of ground, if he is a lad of genius, puts him apprentice either in the mercantile line, or some of the genteel professions. I know this sort of reasoning will draw upon me many enemies; and it will be objected, that by laying a number of small farms together you will depopulate a country. Far be it from me to deprive any man of his property, or to wish to do any thing that may tend to decreafe population; on the contrary, it is my wifh to promote it; convinced that the riches of a country depend upon it. I would not deprive the old farmers of their land; I would have them educate their fons in the ufeful manufactories, and as they die, lay them together, or convert them into manufactories where *properly fituated*, and lay a fufficient portion of land for their convenience; and the reft lay together for the purpofe of farming. Four farms, of 50 acres each, laid together under proper management, would be made to produce one-fourth more for the public market than in feparate allotments; and I think it will be generally confeffed, that, in a country like this, abounding with men of property, ingenuity, and enterprize, that there generally will be found employment in our manufactories for as many inhabitants as there can be found provifions to fupport; confequently the more land is made to produce, the more it will tend to increafe population. I fhall next beg leave to repeat my method of management; which, though you have feen, and I verbally communicated to you, I think may here be more clearly defcribed.

Upon fand land, leamy fand, or dry hazle foils, I cultivate turnips, dreffed with bones, mixed with a portion of fold manure, as communicated to you; next barley, red clover, and wheat; then turnips, barley, white and yellow clovers, pafturing with fheep one or two years; then wheat, and fo on. Upon clay and wet foils, after fallow, wheat, red clover, wheat or oats; then fallow, wheat or barley (if the fallow be limed we always fow wheat; if fold manure, fometimes barley, as I change the tillage as much as poffible), next fmall feeds as above, mixing a few hay feeds, and pafturing with fheep, one, two, or three years, as convenient, or apparently moft ufeful. I then plough out for wheat or oats; if laid more than one year, oats. I have found this, from 30 years experience, to be the most beneficial method of cultivating land; having brought some poor soils to considerable greater value within that period. The farm I occupy is but small, 150 *statute acres*, and, though as well managed 20 years hence as any in the circuit, and as heavily manured, did not then feed more than 20 sheep upon grass, and 40 upon turnip, upon an average. I can now fatten 60, sometimes 80 upon grass, and 100 or 120 upon turnip; and get one-fourth more corn than was formerly raised, besides some increase of other cattle. Here, however, ought to be understood the great expence I am at in artificial manures, these adding to the natural ones in a very considerable proportion. Last year I spread on eighty pounds worth of bones, forty pounds worth of lime, and ten or twelve pounds worth of soot and rape dust, upon this small farm, besides the natural manures it produced: and upon an average it costs me at least L. 100 per year in different sorts of manures. This ought to be considered as a principal means of improvement, and is more by one half than is bought upon an average by the general run of farmers.

I come now to speak of the necessity of leases, which, with the fore-mentioned thoughts on small farms will give answer to your 35th question. The greater part of this county is either tenanted at the will of the proprietor, that is, from year to year, or upon leases clogged with arbitrary clauses, such as being restrained from ploughing out certain pieces of ground under heavy penalties, or confined in some measure to one mode of management, which restrains the genius of the farmer, and ties him from experiments and every useful improvement. There may indeed be a few men found, who will exert their abilities and risk their property under a yearly farm, yet the generality will not: for out of the whole of my acquaintance (and I know a great number of clever farmers), whenever I have asked them, why do you not manage such a piece of land so and so, and how much more would it be made to produce? The answer always is, we are tenants at will, and fear advantage would be taken of our improvements. This, I presume, will appear to every one a natural conclusion. There are two clauses which I think necessary in this country where manure is so dear, and where they are at an inconvenient distance from great towns that manure cannot be replaced, and that is, to be restrained from selling of the hay (*a*) and straw from the premisses; and, four or five years previous to the end of the term, to lay down one-third of the ground in a good husbandry style. These, in my opinion, are all the restraints necessary for the security of the proprietor, and, I think, would not militate against the farmer's interest, but leave him at full liberty to pursue his improvements.

As to the produce of land, good farmers will average from 27 to 30 bushels of wheat per acre, 40 or 44 bushels barley, 64 or 70 bushels oats, and 30 bushels beans. Small farmers and indifferent managers, which occupy, I suppose, 3-fourths of our lands, will not average more than 20 bushels of wheat, 30 bushels barley, 48 bushels oats, and 20 bushels beans. Thus I have communicated to you my opinion upon your different questions to the best of my knowledge, observation, and experience; and where I have erred, it is an error in judgment, which I should be glad to be corrected in.

(*a*) It would be a hardship for a good farmer to be prevented from selling a stack of hay, if he could spare it, in a dear time, when hay rates at L. 4, L. 5, or L. 6 per ton. It would raise him money to buy manure, if it could be had, even at a great distance, and perhaps might leave him more clear profit than he had got by his farm for some years before.

A Yorkshire Farmer.

You are at liberty to make use of my name in any way you think proper: for though it should draw upon me the reproach of the haughty and ignorant farmers, I regard not the censure of such narrow and contracted minds; conscious that it is a duty which every man owes to himself, and mankind, to exert himself for the public welfare, and being convinced that nothing is more necessary, nor can tend more to promote the general interest, than the object you have in view. You have therefore my sincere wishes for its success. I am, &c.

No VI.

EXTRACT from the Correspondence of Mr PARKINSON, at Doncaster.

It is too often little considered how much may be raised from land under good management. It appears to me, that it would be a good scheme for the Board of Agriculture to take a farm into their own hands, and shew, by improved practice, what might be done: this would be of great utility. As to driving any thing into old farmers, it is easier to make new ones. There is land near Doncaster now let at 7s. per acre, which, if managed in a proper manner, and fed by sheep, would pay 20s; and where the sheep that are bred never sell higher than 12s. or 16s. might be fed to 30s. and 40s.

The usual produce per acre, where a rotation of turnips, barley, clover, and wheat, is adopted, is as follows:

Upon poor sands, 3½ quarters of barley, 2 of wheat;—turnip and clover precarious. Upon clay soils, 4 quarters barley, 3 quarters wheat, 3 quarters beans,—clover, and turnips both good. Upon lime-stone, 4 quarters barley, 3 quarters wheat, 2½ quarters beans,—clover and turnips good. Loamy land, 5 quarters barley, 3½ quarters wheat, pease 3 quarters,—clover and turnips good.

The mode of cultivation, however, is very irregular; as the farmers have no leases, they make hay when the fun ſhines, and often crop the ground as long as it will carry. I know a great many farmers who keep their land in a poor ſtate, to prevent the owners from advancing it.

Draining very little known in this part of the country. The wet lands improveable; but the dry lands much more ſo.

With regard to the poor ſands betwixt this place and Bautry, which are at preſent in a very ſhabby ſtate, my opinion is, that the beſt way of going to work with them would be, firſt to begin with a good turnip ſallow, and 10 loads of manure, of 2 tons each to the acre, which may be had at Doncaſter at 5s. per ton, as they have ſcarce any themſelves. This will produce a good crop of turnips, which ought to be eat off with ſheep, and the land ſown with barley and ſeeds—quantity of ſeeds, 1 peck of rye graſs, 14 pound white clover, and 14 pound trefoil. I would paſture it with ſheep for two years, break it for wheat or rye, and return to turnips. My reaſons for this are; rye graſs is a very good winter plant, and ſcarce can be eaten too near in the ſpring, when graſs is of moſt value. If it run to a bent, it exhauſts itſelf for that ſeaſon, and is worth nothing till autumn. Trefoil is more early than white clover; therefore, with theſe mixtures, three different ſprings are got. Many farmers like red clover; I do not, except for cutting and I think it much the better of a little rye graſs. Red clover, on many ſoils, ſtands but for one year, therefore is very improper ſeed for paſture, which thoſe ſands ſhould be applied to as much as poſſible, to faſten them. All artificial graſſes ſhould bear two years eating at leaſt, the expence of ſeeds being great; but none will ſcarcely bear more than three years. No poor ſand or lime-ſtone ought to be paſtured longer than it will keep a ſufficient number of ſheep to leave a good top dreſſing when ploughed up; by reaſon the land is then loſing what was put into it before, and returning to its natural ſtate. In time, a hot bed will come to earth. Manure, mixed with ſoil, cauſes fermentation in ſome meaſure, like yeſt put amongſt wort, and will ſoon go off, and ceaſe to operate.

The land, in its preſent ſtate of cultivation, lets high; though worth double the ſum if properly managed. Sheep are much wanted, as there is no improvement equal to the ſheep farming: it is both the cheapeſt and beſt upon all dry ſoils. If the farmer could only be made to underſtand he had a ſort of inheritance in his farm, which can no way be done but by giving leaſes, it would be of general utility to the kingdom at large. The farmers are the firſt and the grand machine of all improvements, and therefore ought to have every poſſible encouragement given them. I never was in any part of the country where the people were more ſlat to improvement than in this neighbourhood. I apprehend the cauſe is this, a great many gentleman live in it, conſequently near their tenants, and are curbs upon their ingenuity. Moſt experiments are coſtly, and the farmer is affraid his landlord will look upon his attempts to improve as acts of extravagence,—ſuch as hiring a Diſhley ram for 100 guineas the ſeaſon, and other things of the ſame kind.

There is an abſurd idea ſome men have, that the ſcheme I have adopted for the ſands will diminiſh the quantity of grain: I ſay no,—it will only add to it; for an acre managed in the way I have deſcribed, will produce as much as two do now. As for the ſmall mutton and fine wool that would be loſt by my ſcheme, there will always be plenty of the former on the mountains, for the tables of the great; and if lambs are clipped, they will produce fine carding wool, which does away theſe objections.

No VII.

STATE of the WASTE LANDS in Yorkshire, calculated by MR TUKE, Junior.

	Capable of cultivation, or of being converted into Pasture.	Incapable of being improved except by planting.	Total.
Waste lands in the *North Riding*.	Acres.	Acres.	Acres.
The Western moor lands —	150,000	76,940	226,940
Eastern ditto — —	60,000	136,625	196,625
Detached moors, or waste, in the country — —	18,435	— —	18,435
Total —	228,435	213,565	442,000
Waste lands in the *West Riding*.			
The high moors — —	200,000	140,272	340,272
Detached moors, or waste, in the country — —	65,000	— —	65,000
Total —	265,000	140,272	405,272
Waste lands in the *East Riding*			
Detached moors, or waste, in the country — —	2,000	— —	2,000

In the North Riding — 442,000
West Riding — 405,272
East Riding — 2,000

Total waste lands in Yorkshire — 849,272

No VIII.

OBSERVATIONS by Mr Day of Doncaſter, regarding the Size of Live Stock.

I am much inclined to believe, that breeders in general, are deſirous of breeding their cattle of too great a ſize, which is neither for their own advantage, nor for that of the country in general. My opinion is, that oxen weighing from 40 to 60 ſtone, are the moſt uſeful to the conſumer, and worth more per ſtone than greater weights. There are other advantages attending ſmall cattle. There are many parts of England, where the land would juſt ſupport cattle of from 80 to 90 ſtones, that would fatten, and conſequently would bring to perfection, thoſe of from 40 to 50 ſtone. This plainly ſhews that middling weights, are the moſt generally convenient, and conſequently the moſt profitable to the grazier. Nor can I believe, that the ſmaller weights are ſo liable to diſeaſes, being in general hardier; but if they ſhould happen to die, the loſs of an ox of 40 ſtone weight is not ſo much felt as one of a larger ſize. Smaller animals alſo, are in general quicker feeders, where the ſhape of the animal is attended to. There is no ſort of breed, that on the whole, I am fonder of, than the Galloway ſcot, as the beef is of very good quality, and their ſize is well calculated for general conſumption. I beg leave to add, that of all the ſigns of a good feeder, there is none I prefer to that of having a ſmall head. It is rare indeed to ſee a large coarſe headed animal a good thriver.

In regard to sheep, my opinion is the same; namely, that sheep, which, when fat, will weigh from 14 to 20 lb. per quarter, are proportionably of more value than those which weigh from 20 to 30 lb. There seems to me not the least doubt, that the smaller, in this case, is preferable to the larger animal: For instance, six sheep, at 16 lb. per quarter, equal, in point of weight, to four at 24 lb. per quarter; but if it can be proved by experiment, which any one may soon do to his complete satisfaction, that the six sheep would fatten sooner, and on less land than the four, can there be any doubt which is the best sort for the individual and for the public? Besides, the risk of less loss by the death of the smaller animals, is here also an object worthy of attention.

In regard to the wool, it is much more than probable, that the fleeces of the six smaller sheep will be more valuable than those of four of the larger sort.

On the whole, I am of opinion, that the smaller sorts of live stock, are preferable to the larger, and that the arguments in their favour, ought to be as generally known as possible, both among breeders and graziers, and indeed to the public at large, in order that any tendency for breeding the unprofitable larger stocks may be checked as much as possible.

No IX.

ACCOUNT of the different TOWN-SHIPS in the Wapentake of CLARO, from materials furniſhed by ROBERT STOCKDALE, Eſq.

Humburton with *Milby*—The greateſt part of this townſhip is the property of Jacob Smith, Eſq; and occupied by him. He is improving it by banking out the floods, ploughing the old paſtures, and draining them completely. Where the land is rough and ſour he invariably pares and burns; and as he is an active intelligent farmer, the huſbandry practiſed in this townſhip is of the moſt perfect kind.

Aldbrough—This townſhip is not very extenſive, and conſiſts of open fields, and incloſed graſs land, nearly in equal portions. Turnips, barley, clover, and wheat is the uſual rotation, and the land is well managed. A ſmall common, of about 150 acres, belongs to the townſhip.

Boroughbridge—There are only about 30 acres of incloſed garths and crofts, and a ſmall common of 60 acres belonging to this townſhip, which appertain to the borough houſes.

Minſkip—Is nearly under the ſame circumſtances as Aldbrough in reſpect to ſoil and cultivation, being moſtly occupied by ſmall farmers, and ſmall freeholders; has no common.

Rackliffe—The ſoil of this townſhip is moſtly a ſtrong clay, and the rotation is, *1ſt*, fallow; *2d*, wheat; *3d*, oats or beans, and fallow again. The farms are all ſmall, but well managed.

Stavely—There is a ſmall common here not exceeding 50 acres. The cultivation, &c. ſimilar to the townſhips of Aldbrough and Minſkip.

Copgrove—This townſhip is principally, if not wholly, the property of Henry Duncomb, Eſq; and conſiſts of a variety of ſoils, and cultivated in various ways. Some farms are almoſt wholly arable, but in general they are a mixture of arable, paſture, and meadow land, like the reſt of the Wapentake.

Burton Leonard—This townſhip has lately been incloſed under an act of parliament; and where the turnip huſbandry can be purſued, it is adopted.

Nidd.—Is all incloſed, and kept nearly in equal proportions of corn and graſs. Farms are of ſmall ſize.

Stainley—Is nearly in the ſame ſtate as the townſhip of Nidd.

Brearton—Has been lately incloſed, and differs little as to ſize of farms and cultivation from the two laſt mentioned townſhips.

Scotton—Has a common of 200 acres of good land,

and the remainder of the land is in almost equal divisions of open fields and inclosures.

Farnhap—This township is wholly inclosed. No common.

Arkendale—This township lately consisted of a common, and open fields, but they are now divided and inclosed by an act of parliament.

Knaresbrough—The land around this town is chiefly in grass, and occupied by the inhabitants, who are mostly manufacturers of linen, which is carried on to a great extent. The only part let out in farms is an estate left for the support of dissenting ministers.

Scriven—A common of 200 acres of rich land belongs to this township.

Allerton with *Flaxby*—Is wholly the property of Thomas Thornton, Esq; and is let out in small farms.

Goldsbrough—Belongs to Lord Harewood, and consists of a wet swampy common of 400 acres, and the residue mostly in open arable fields. The farms small.

Ribstane little—This township is exactly under the same circumstances as Goldsbrough, except that the common is good land, and under a regulated stint.

Plimpton—The soil varies much, and of course, the rotations of cropping are different. The farms are rather larger than in the neighbourhood, and kept in a high state of cultivation.

Spofforth—Has been lately incloſed, and conſequently is in an improving ſtate.

North Deighton—Exactly under the ſame circumſtances as to ſoil, diviſion, and cultivation, as the laſt.

Kirk Deighton—Is wholly incloſed, and chiefly occupied by ſmall freeholders.

Midleton with *Stockeld*—Is wholly the property of Mr Midleton a Roman Catholic, and is let in farms rather larger than uſual there. Stockeld is all incloſed, but Midleton, which is ſituated at 20 miles diſtance, and lies nearly at the moſt ſouth-weſt point of the Wapentake, has a large extent of mountainous heathy common, conſiſting of 1500 acres at leaſt.

Linton—This townſhip is moſtly incloſed and let in ſmall farms.

Wetherby—The land here belongs chiefly to the Duke of Devonſhire, and is incloſed and let out in ſmall farms. The turnip huſbandry practiſed, where it can be done with advantage.

Coutkrop—Belongs to Lord Petre, and is moſtly in open fields. The farms are ſmall. It is in this townſhip that the large oak tree ſtands, which is deſcribed in Dr Hunter's *Evelyn's Sylva*.

Hunſingore, *Kibſtoon*, and *Cattalgreat*—Theſe three townſhips are almoſt the excluſive property of Sir Henry Goodrick, Bart. Hunſingore and Cattalgreat lately conſiſted of open arable fields, but are now incloſed.

One third meadow, one third paſture, is the beſt mode of managing a farm in theſe lands.

Whixley—This townſhip principally belongs to a charitable eſtabliſhment for 12 decayed gentlemen and 12 ſtudents at Cambridge. It conſiſts moſtly of open arable fields, two ſtinted paſtures, and an unſtinted wet common of about 100 acres, alſo 1000 acres of incloſed land let in ſmall farms.

Thornville or *Little Cattal*—This is a ſmall diſtrict belonging moſtly to Thomas Thornton, Eſq; and is chiefly in rich paſtures.

Kirk Hammerton—This townſhip has been lately incloſed by an act of parliament.

Nun Monkton—This is a ſmall townſhip, the property of William Tuſſnel Joliffe, Eſq. The ſoil is ſtrong clay, and the rotation uſually practiſed is, *1ſt*, fallow; *2d*, wheat; *3d*, beans or oats.

Low Dunsforth, *Green Hammerton*, and *Marton*, with *Grafton*—Turnips are generally cultivated upon the tillage lands of theſe townſhips, and the farms are ſmall. Grafton has a common of about 100 acres of good land.

High Dunsforth—Has lately been incloſed, and turnips are cultivated where the ſoil is proper for that root.

Great Ouſeborne Pariſh—Has alſo been lately incloſed. At Branton, in this pariſh, reſides the beſt farmer in the Wapentake. He had long ago adopted the turnip and clover huſbandry, and when he found his land tire of clover, he then ſowed beans in drills, which he followed with barley, wheat, and turnips. He then discovered, that after a repetition of beans, neither his wheat nor his turnips were so good, therefore sows white clover, trefoil, &c. which he eats with sheep, and his wheat and turnips both flourish.

Little Ouseborn—This township consists of open fields, and inclosed arable land, with a small stinted common, not exceeding 70 acres, and being mostly a light dry soil, has been long under the turnip husbandry; but the continued succession of the same routine of crops now proves to be injurious, as red clover will scarce grow at all.

Kirby-Hall—Is a small township belonging to Henry Thomson, Esq; and consists of meadow and pasture. He occupies a considerable part of it himself, and is very attentive to the breed of cattle and sheep.

Sickling Hall, *Kirby Overblow*, and *Keerby*, with *Netherby*—These townships consist both of open arable fields, and inclosed arable and grass lands. The soil varies, and of course the husbandry. The turnip-husbandry is not much practised. Each of these townships has betwixt 200 and 300 acres of common, which might easily be divided and inclosed under one act of parliament, as they are contiguous.

Rigton—This township had an extensive common of 2000 acres, which was inclosed by act of parliament in 1775, and is now nearly in equal portions of arable and grass. Few turnips are grown, but the tenants take wheat after fallow, and then beans or oats This inclosure, without the addition of any manufacture, has increased the number of inhabitants as two to one in eight years. It has alſo increaſed the annualrent of the townſhip above double, and many parts of it will yet admit of very great improvements.

Ripley—This townſhip conſiſts of ancient incloſures, which are moſtly kept in graſs. The arable land is employed in raiſing turnips and potatoes, with a ſucceſſion of barley, clover, and wheat.

Thornton and *Scarrow*—Theſe two townſhips are in the pariſh of Ripley, which is a rectory, in the patronage of Sir John Ingleby, Baronet; and though incloſure acts have been obtained for the commons of both places, yet the property is ſtill tithable.

Markington—Conſiſts moſtly of ancient incloſures, which are kept in nearly equal proportions of arable and graſs. Turnips are ſown where the ſoil is proper for them.

Folliſoot—Has been lately incloſed, and about 1500 acres of common brought into cultivation. The ſoil is in general ſteril, and in ſome places too ſtony for the plough.

Dunkerwick, *Weeton*, and *Huby*—Theſe three may be taken together. Their commons, conſiſting of about 1000 acres of rich land, have been lately divided and incloſed. The general mode of breaking up commons or old graſs incloſures here is, 1ſt, to pare and burn; 2d, to take rape or turnips; afterwards wheat or oats: Beans run too much to hulm or ſtraw in freſh land.

Stainburn—This townſhip has a number of ſmall incloſures, which appear to have been gradually taken off the common by the cottagers, but the common ſtill contains above 2000 acres of valuable land, capable of great improvement; and though an incloſure act was obtained fourteen or fifteen years ago, yet nothing more has been done than to ſet off an allotment in lieu of tithe.

Caſtley and *Leathley*—Are moſtly in graſs, being rich feeding lands, adjoining the river Wharfe.

Linley—Conſiſts of ſmall ancient incloſures, with an extenſive common of at leaſt 1000 acres of tolerable land.

Farnley—Has an extenſive common belonging to it, which the lord of the manor, being ſole proprietor, is gradually improving, by partial incloſures, and plowing. The common that remains is nearly 500 acres.

Newall with *Cliſton*—Conſiſts of rich incloſures, moſtly in graſs. The common belonging to it was incloſed about twelve years ago.

Weſton and *Aſkwith*—Theſe townſhips conſiſt chiefly of rich paſture and meadow land. Weſton has a ſmall common, not exceeding 100 acres. The common of Aſkwith, containing about 1200 acres, was, by virtue of an act of parliament, about twelve years ago, aſſigned to the impropriate rector, Walter Vavaſour, Eſq; in lieu of tithes, which he is gradually converting into ſmall farms; but many parts of it are only fit for planting, being too rocky for the plough.

Denton—The low part next the river Wharfe is rich paſturage. There is, however, a good deal of arable land, and a common of perhaps 1500 acres.

Nesfield and *Langbar*—Nearly the fame as Denton, with the like quantity of common.

Haverach Park—This townfhip is extra-parochial, and was formerly a park belonging to the foreft of Knaref-borough.

Timble Little—Is a fmall townfhip, moftly in grafs, with a common of about 100 acres of good land.

Beamfly with *Hazlewood*, and *Hartwith* with *Winfley*—Thefe townfhips are of great extent, and moftly kept in grafs. The commons annexed to the firft contain at leaft 2000 acres, and the laft about 1500 acres.

Dacre—Much in the fame fituation as the two laft. The commons contain about 2000 acres, one half of of which is ftinted pafture.

Baverley—Moftly kept in grafs, and has a large extent of wafte land, which is replete with coal and lime. Many mines of lead ore are now working to advantage. The common contains about 3000 acres.

Pately Bridge—The land on each fide of the river Nid abounds with fprings, which are turned to great advantage in bleaching linen yarn and cloth, the principal manufactures of this town and neighbourhood. The land is therefore principally in grafs, and let in fmall parcels at an average of 40s. per acre. The mountains produce lead, and the herbage is of a coarfe nature called *bent*. Thefe mountains are of confiderable extent, and are ufed chiefly, though not wholly, as ftinted paftures.

Fountaines-Earth, *Stonebeck-up*, *Stonebeck-down*—Thefe

three townſhips are ſituated in the midſt of high moors. Their incloſed fields are moſtly kept in graſs, with a ſmall portion of arable land. The commons are very extenſive, at leaſt 5000 acres.

Kirkey Malzeard—This townſhip conſiſts of ancient ſmall incloſures, moſtly kept in graſs. What part of it is kept in tillage, is ſown with turnips, where the ſoil admits. The waſtes are extenſive, but the number of acres not aſcertained.

Azerley, Laverton, Studley Roger and *Studley Royal*—Theſe four townſhips conſiſt of rich paſturage, and are uſed as dairy farms. Turnips are ſown on the tillage lands where the ſoil anſwers, and thoſe of a different nature are cleaned, by a plain ſummer fallow.

Sawley, Grantley and *Aldfield*—Theſe townſhips are moſtly in graſs. They have extenſive commons pertaining to them, of at leaſt 1000 acres.

Ripon—The land ſurrounding this beautiful town is moſtly in graſs, and occupied in ſmall parcels by different tradeſmen, &c. reſiding there.

Little Thorp and *Biſhop Monkton*—The land in theſe townſhips conſiſt both of ſmall incloſures of graſs, and open arable fields. In ſoils adapted thereto, turnips, with the uſual conſequent crops, are ſown; and in thoſe of a different nature, fallow, with its cuſtomary rotation, is practiſed.

No VIII.

STATISTICAL INFORMATION concerning different PARISHES.

COLLECTED ſtatement of intelligence received by the Deputy Clerk of the Peace for the Weſt Riding of Yorkſhire, in anſwer to queſtions tranſmitted by him to the Miniſters of the reſpective pariſhes in ſaid Riding, by direction of the Magiſtrates appointed to correſpond with the Board of Agriculture. Tranſmitted by the Right Honourable Lord Hawke.

Pariſh of *Ackworth* contains,

1242 Inhabitants
2442 Acres of ground as *per* ſurvey
1431 Acres of graſs eſtimated
1011 Ditto arable ditto.

Rotation of crops—Fallow
Wheat
Oats
Beans
or
Turnips
Barley, or Oats
Wheat.

Pariſh of *Addle*,

958 Inhabitants
6660 Acres of gronnd
1418 Ditto of graſs

4255 Acres arable
666 Ditto waſte
331 Ditto woods.

Crops for one year,

Fallow, Clover, and Turnips	1063 acres
Barley	1000 do.
Oats	1300 do.
Wheat	800 do.
Beans	92 do.

This account is given not as being accurate, but as near as the writer could calculate.

Half of the waſte is improveable.

The ſame quantity of land is yearly in the ſame mode of cultivation.

Pariſh of *Addlingfleet*,

341 Inhabitants
5000 Acres of ground
Two-thirds of the above in graſs, and the other one-third arable
1000 Acres a waſte.

Crops—One-fifteenth flax
One-fifteenth rape and turnip
One-tenth potatoes
The reſt grain.

This is a narrow ſlip of land, eight miles long and one broad.

Its waſtes conſiſt of an undivided moiety of thorn moors, ſome part of it valuable, but the major part uſeleſs for want of proper drainage; and, as far as the writer knows, a conſiderable drainage would be difficult, the parish being situated below the level of the sea at high water, and consequently is marshy. The land is a rich strong loam, very favourable to grazing occupiers, some of whom rent sixteen or eighteen hundred acres each, which, the writer observes, is the cause why the parish is so thinly inhabited. An act for a partial inclosure was obtained in 1769, and great improvements in consequence are made in the parish. Many principal persons were desirous, in 1795, of extending the inclosures, though no hope was entertained of rendering the whole waste valuable.

Parish of *Adwick*,

280 Inhabitants
1416 Acres of ground
300 Acres of grass
1116 Ditto arable.

The usual course of tillage in the part of the country.

Parish of *Armcliffe*,

900 Inhabitants
25000 Acres of ground
7000 Ditto grass
190 Ditto arable
17900 Ditto waste.

Of the 7000 acres in grass the whole is inclosed and divided, 3000 are cut for hay, and 4000 are in pasture. Of the 17900 acres in a state of waste, the greater part is barren and mountainous, but divided into cattle gates, the number of which is only known to the proprietors.

Parifh of *Armthorpe*,

50 Families
2000 Acres of ground
700 Ditto grafs
1300 Arable.
Rotation of crops—1ft Turnips
2d Barley two-thirds and oats one-third
3d Wheat and rye
4th Clover, Fallow.

N. B. It is fuppofed the writer has made a miftake in placing the wheat as fown on the barley ftubble, and having the clover fown with it, inftead of the clover being fown with the barley.

The houfes of every defcription, including cottages, are very few, of any confideration; the largeft containing about 12 people including all the fervants.

Parifh of *Bardfey*,

300 Inhabitants
3000 Acres of ground
30 or 40 Acres of wafte
and 300 Acres in rabbit warrens.

The greateft part is arable. A very fmall quantity in grafs.

As the parifh boundaries, on all fides, are difputed and cannot be afcertained, the number of acres in the parifh muft alfo be uncertain.

A very fmall quantity of land is in grafs; for the foil is of that nature as to require being plowed out every third or fourth year, letting it remain longer in pafture being found unprofitable to the occupier.

The wafte land is not worth cultivation. The 300

acres in rabbit warrens is ordered by the proprietor to be converted immediately into arable land.

Parish of *Barnbydunn*,

536 Inhabitants
3192 Acres of ground
841 Acres of grass
1851 Ditto arable
500 Ditto waste
1392 Ditto in corn and clover annually
459 Ditto fallow.

This parish contains three townships. 1st, Barnbydunn; 2d, South Bramonth; 3d, Thorpe; and in each township the cultivation is as under,

	In the First	Second
	Acres.	Acres.
Wheat	183	52
Clover	183	52
Barley and oats	367	105
Fallow	183	52
Grass	565	52
Waste	500	—

The writer's return for the 3d township is,

Arable land	448 Acres
Fallow	224 Ditto
Grass	224 Ditto.

Parish of *Batley*, divided into 4 townships, as under,

1. 1576 Inhabitants
1599 a. 2 r. 27 p. of ground
1031 a. 0 r. 29 p. grass
525 a. ——— arable
42 a. 3 r. 11 p. waste
421 a. 2 r. 27 p. in corn
104 a. ——— fallow.

2. 489 Inhabitants
408 Acres of ground
258½ Ditto grafs
149½ Ditto arable
98½ Ditto in corn
51 Ditto fallow.

3. 1800 Inhabitants
700 Acres of ground
600 Ditto grafs
109 Ditto arable
100 Ditto corn and fallow.

4. 1801 Inhabitants
2321 a. 2 r. of ground
1144 a. 2 r. grafs
1117 a. 2 r. arable
59 a. 2 r. wafte
789½ a. in corn
328 a. fallow.

This parifh contains 4 townfhips. 1ft, Batley; 2d, Churchwell; 3d, Gildirfome; 4th, Mofley.

Parifh of *Gargrave*,

From 800 to 900 Inhabitants

10000 Acres of ground by eftimation.

This parifh contains fix townfhips, and the writer fays, that, by eftimation, they contain 10000 acres, of which not one hundred acres is plowed, nor twenty acres wafte, all the parifh being entirely grazed.

Parifh of *Garforth*,

500 Inhabitants
1440 Acres of ground by eftimation
547 Acres of grafs

643 Acres arable
250 Acres waſte

Of the 547 acres of graſs, 282 acres are meadow, and 265 acres paſture.

Pariſh of *Giggleswick*,

2200 Inhabitants
16500 Acres by eſtimation
14685 Acres graſs
315 Acres arable
150 Acres waſte
300 Acres in oats
15 Acres barley

500 acres are occaſionally plowed in ſmall quantities.

Pariſh of *Guiſeley*,

This pariſh contains 5 townſhips as under, viz.

1. Guiſeley	771	Inhabitants
	964	Acres of ground
	678	Acres graſs
	286	Acres arable.
2. Carleton	104	Inhabitants
	1181	Acres of ground
	751	Acres of graſs
	430	Acres arable.
3. Yeaden	1517	Inhabitants
	1080	Acres of ground
	861	Acres graſs
	218	Acres arable.
4. Horſeforth	2230	Inhabitants
	2226	Acres of ground
	1369½	Acres graſs
	856½	Acres arable.

5. Rawden, no return made of this townfhip.

	First	Second	Third	Fourth
	a.	a.	a.	a.
Wheat	44	52	46	—
Barley	38	38	27	—
Potatoes	26	23	11	—
Oats	111	175	68	583½
Beans	1	4	14	—
Turnips	3	19	14	—
Fallow	63	119	39	273

Parifh of *Hatfield*,

2000 Inhabitants
8830 Acres of ground by eftimation
3858 Acres grafs
4972 Acres arable

	Acres.
Wheat and rye	1180
Oats	1145
Beans and peafe	75
Fallow	1151
Clover	762
Barley	592
Potatoes	67.

Perhaps 300 acres of the fallow is fown with turnips.

This letter-writer very properly finds fault with the population of villages, by throwing them into large grafs farms, and the impolicy of landlords, not letting their tenants occafionally plow even a fmall part of their farm. He ftates, from authority, a refufal to a farmer to plow as much ground as would fupport his own family with bread, and his ftable with ftraw. He fays, that he travelled lately through Craven, where, whilft immenfe fums of money are expended in inclofing various parts, in others many inclosures are thrown into one, and occupied by one farmer only, tho' before by many. He says, that it would be no difficult matter to prove, that the commons of Hatfield, Thorn, and Frishlake, under proper regulation and flint, would be more useful, singularly and publicly than when inclosed*. That these commons cannot properly be called waste, as they support many thousands of stock, and the inhabitants have an equal stray upon them. He also finds fault with large farms in general, and mentions the depopulation of villages in consequence of them.

Parish of *Hampsthwaite*,
from 1700 to 1800 Inhabitants
12800 Acres of ground
6000 Acres of grass
3000 Acres arable
3800 Acres waste.

The cultivation is 2000 acres in different kinds of corn, but chiefly oats, and 1000 acres fallow.

The letter-writer says, that the parish is 8 miles long, and at a medium two miles and a half wide; containing 20 square miles, or 12800 acres; and that the 3800, entered as waste, though inclosed, can be called nothing but waste, as it has not been cultivated, and the owner of a part of it living at a distance, would sell 1000 acres of it.

Note by the Editor.

* Without insisting upon the utility of inclosing these commons, it certainly must prove highly advantageous that they were divided. As to what is said concerning the depopulation of villages in consequence of large farms, these ideas are wholly unfounded; for, whether the farm is large or small, if the management is similar, the number of people on a given quantity of land, will in all times nearly be equal.

Pariſh of *Handſworth*,

1423 Inhabitants
3000 Acres of ground by eſtimation
400 Acres waſte.

The letter-writer ſays, that the 400 Acres of waſte is about to be incloſed.

Pariſh of *High Hegland*,

This pariſh contains three townſhips viz.

	a.	r.	p.	
1. High Hogland	730	2	8	by ſurvey
	243	2	2	graſs
	487	0	6	arable
	9	0	0	waſte
	100	0	0	woods.
2. Clayton	1378	1	38	by ſurvey

one half graſs, and one half arable
200 Acres waſte
32½ Acres woods.

3. Skilmorthorp 595 Acres per ſurvey
one-third meadow and paſture, and two-thirds arable,
80 Acres waſte
80 Acres woods.

No return of the inhabitants in this pariſh; and the letter-writer ſays, that the 9 acres of waſte in No 1 is not worth incloſing; that the 200 acres in No 2d, if incloſed, would make good corn or graſs land; and that the 80 acres in No 3d, is well worth the incloſing.

Pariſh of *Hooton Roberts*,

148 Inhabitants in 30 houſes
1015 Acres by ſurvey
550 Acres graſs and paſture
410 Acres arable
55 woods and roads
8 ſmall common.

Cultivation,	160 Acres wheat
	60 Acres barley
	90 Acres oats
	30 Acres beans and peafe
	120 Acres fallow
	50 Acres clover.

Parifh of *Horton in Ribblefdale*,

663	Inhabitants
17280	Acres by eftimation
7360	Acres grafs
2560	Acres arable
7360	Acres wafte
24	Acres oats.

The letter-writer fays, this account is far from being accurate, but is made from the very beft information that he could obtain.

Parifh of *Ilkley*,

This parifh contains 3 townfhips.

1. Ilkley	109 families or 545 inhabitants
	1800 Acres inclofed land
	1379 Acres meadow or potatoes
	371 Acres arable
	2400 Acres of moor or common
	50 Acres of woods
	95 Acres fallow, turnips or potatoes.
2. Middleton	42 families or 168 fouls
	716 Acres inclofed land
	466 Acres meadow or pafture
	150 Acres arable
	450 Acres common
	100 Acres woods.
3. Nefsfield, cum Langbar	46 families or 230 fouls

923 Acres inclofed land
200 Acres arable
200 Acres common
73 Acres woods.

This letter-writer fays, that the ufual method of cultivation is two crops, then a fallow, then a fingle crop and grafs feeds, 3 lb. of clover to 2 lb. of treyfoil, with 2 quarters of hay feeds, which hold good for two years, but no longer. The two laft townfhips are tithe-free.

Parifh of *Killington*,

This parifh contains 4 townfhips, viz.

1. Killington	1193 Acres inclofed land
	360 Acres meadow
	833 Acres arable.
2. Beaghall	1600 Acres inclofed land
	600 Acres pafture
	1000 Acres arable.
3. Egbrough	1800 Acres
	600 Acres pafture
	1200 Acres arable
	300 Acres wafte
	1800 Acres open fields
4. Whitley	1300 Acres
	900 Acres pafture.

The letter-writer fays, that the 300 acres of wafte in No 3, if inclofed, are capable of great improvement, as alfo the open field of 1800 acres; and that the land of this parifh has been improved within the laft 30 years, from 5s. to 20s. per acre, and the rectory from

L. 180 per annum to L. 600. The population much the fame as 20 years ago. A great deal of land fown with feeds, and eat off with fheep. Fallows always fown with turnips, then barley, clover, and hard corn. Whitely was inclofed in the year 1774; Killington and Beaghall about two years ago; Egbrough ftill uninclofed.

Parifh of *Kirk Bramwith*,

	249	Inhabitants
	1712	Acres per furvey
	517½	Acres grafs
	1194½	Acres arable.
Cultivation,	45	Acres barley
	300½	Acres wheat or meflin
	306½	Acres oats
	225	Acres beans
	57½	Acres clover
	21	Acres flax
	14	Acres turnips
	12½	Acres potatoes
	231½	Acres fallow.

This parifh being fubject to frequent inundations, no regular courfe can be followed, as fome farmers fow three times for one crop. The reafon why the quantity of fallow appears fmall, is becaufe of the farmers having land in two parifhes, fome of their land will fall fometimes in one parifh and fometimes in another.

Parifh of *Kirk Heaton*,

1053	families
4060	Acres inclofed, by eftimation
500	Acres wafte.

The letter-writer fays, that he can give no certain account of the quantity of corn grown in this pariſh, only that there are but five farmers in it, as the land is let off in ſmall quantities for the accomodation of trade, and conſequently little corn grown. Perhaps one-third may be wheat, another oats, another beans, barley, potatoes or turnips, as convenience requires.

Pariſh of *Kirkſmeaton*,

231 Inhabitants
1419 Acres of land
284 Acres graſs
855 Acres arable
280 Acres waſte
230 Acres wheat
180 Acres barley
95 Acres oats
70 Acres clover
50 Acres beans
230 Acres fallow.

The letter-writer ſays, that of the 280 acres of waſte, 200 is common, and 80 acres town paſture, and that the town paſture, on account of its ſituation, is utterly incapable of cultivation.

Pariſh of *Long Preſton*,

1299 Inhabitants
7141 Acres of land
4132 Acres meadow or paſture
1983 Acres arable
1626 Acres moor or ſtinted paſture.

The principal crop in this pariſh is oats, ſome turnips and a little wheat are alſo ſown.

Pariſh of *Marr*,

154 Inhabitants
1750 Acres per ſurvey
223 Acres graſs
1380 Acres arable
147 Acres woods
500 Acres wheat
620 Acres barley oats or beans
460 Acres fallow.

Pariſh of *Marton*,

240 Inhabitants
1383 Acres Yorkſhire cuſtomary meaſure, being by a chain of 28 yards, inſtead of 22 yards.

The letter-writer ſays, that the meaſure is taken from an old regular ſurvey land-tax book: That the arable land in the pariſh is greatly decreaſed within the laſt 20 years, nearly the whole being in graſs for feeding cattle, and that this year there is no arable land, ſave about 7 acres of oats in ſmall incloſures. He adds, that there is very little waſte land.

Pariſh of *Mirfield*,

600 Families
3200 Acres of land
2095 Acres graſs
655 Acres arable
455 Acres waſte
250 Acres woods
280 Acres wheat
25 Acres barley
215 Acres oats

25 Acres beans
30 Acres turnip
80 Acres fallow.

It would seem by the number of acres, compared with the 600 stated by the letter writer as families, that he must mean souls.

Parish of *Conisbrough*,

840 Inhabitants
4200 Acres by estimation
1450 Acres grass
2290 Acres arable
460 Acres waste
730 Acres wheat and rye
235 Acres barley
260 Acres oats.

The greater part of the waste land is capable of improvement. The population has considerably increased within the last 20 years, it amounted to 840 souls, November 1795. It is increased in a greater proportion than the crops within the same period. The common fields have been little improved. A considerable part of the parish is inclosed. Three crops on a fallow is a general course, except on inclosed farms, where frequently only two crops are taken. The number of acres in corn, is uncertain, sometimes more, sometimes less.

Parish of *Crofton*,

524 Inhabitants
1340 Acres
500 Acres grass
2 Acres waste
277 Acres wheat

90 Acres oats
52 Acres barley
38 Acres beans
167 Acres clover
214 Acres fallow.

Parish of *Darton*,

1300 Inhabitants
3240 Acres estimated
1500 Acres grass
1500 Acres arable
240 Acres waste
500 Acres turnip
360 Acres barley
500 Acres wheat
3000 Acres inclosed.

Parish of *Dewsbury*,

1040 Inhabitants
1533 Acres per survey
one-third of which is grass, and two-thirds arable
187 Acres woods.

Cultivation, one-fourth wheat
one-fourth beans and barley
one-fourth oats and clover
one-fourth fallow.

Parish of *East Ardsley*,

610 Inhabitants
1581 Acres per survey
690½ Acres grass
699 Acres arable
120 Acres waste
70½ Acres woods
284 Acres wheat

130 Acres oats
120 Acres fallow
70 Acres barley
40 Acres beans
40 Acres turnips
10 Acres peafe
5 Acres potatoes.

Under the head of grafs, the letter-writer includes every kind of grafs, clover, &c.

Parifh of *Edlington*,

110 Inhabitants nearly
1593½ Acres from furvey
one-fourth of which in grafs
300 Acres woods.

The letter-writer fays, that one-fourth is grafs, one-fourth fallow, one-fourth wheat or barley, and one-fourth peafe, beans or oats. Sometimes two and fometimes three crops to a fallow.

Parifh of *Emley*,

This parifh contains two townfhips, Emley and Skilmanthorp.

Inhabitants in Emley 1117
Inhabitants in Skilmanthorp 525

Amount 1642 Inhabitants
3420 Acres from furvey
1275 Acres grafs
400 Acres wafte
318 Acres fallow
416 Acres wheat
597 Acres oats
42 Acres beans

54 Acres peafe
96 Acres barley
104 Acres clover
22 Acres potatoes
96 Acres turnips.

Parifh of *Ferrybridge*,

This parifh contains 3 townfhips,

1. Ferrybridge 200 Inhabitants
1056 Acres from furvey.

2. Waterpyftone 2000 Acres from ditto.

3. Wildon 600 Acres from ditto
120 Inhabitants in two laft townfhips.

Parifh of *Fifblake*,

1078 Inhabitants
3992 Acres eftimated
3992 Acres arable
193 Acres wafte
854 Acres wheat
530 Acres oats
446 Acres beans
55 Acres barley
29 Acres potatoes
20 Acres turnips
125 Acres clover
48 Acres flax
700 Acres fallow
1185 Acres grafs.

The letter-writer fays, that upon the inclofing of the wafte lands in this parifh, the proprietors of certain meſſuages in the pariſh of Fiſhlake, would be entitled to 1832 acres of ſuch incloſure in the manor of beſides the 193 acres of waſte above ſtated.

Pariſh of *Frickley cum Clayton*,

400	Inhabitants
1850	Acres eſtimated
500	—— graſs
1000	—— arable
350	—— waſte
350	—— wheat
100	—— beans
150	—— oats
45	—— turnips
5	—— potatoes
75	—— barley
75	—— clover
200	—— fallow.

Pariſh of *Felkirk*,

310	Men
329	women
332	children
5495	Acres eſtimated
3344	—— graſs
1396	—— arable
255	—— waſte
350	—— fallow
150	—— clover
896	—— corn.

Pariſh of *Slaidburn*,

360	Inhabitants
23950	Acres of land eſtimated
10100	—— graſs

1850 Acres arable
18000 —— waste
850 —— oats.

The letter-writer says, there are nothing but oats grown in this parish.

Parish of *Tadwick*,

170 Inhabitants
1700 Acres estimated
734 —— grass
966 —— arable
337 —— wheat
271 —— oats
107 —— barley
250 —— fallow.

Parish of *Thorne*,

2000 Inhabitants
6086 Acres estimated
1936 —— grass
4150 —— arable
1000 —— wheat and rye
850 —— fallow
1000 —— oats
150 —— beans, &c.
700 —— clover
300 —— barley
150 —— potatoes.

About 150 acres of the fallow sown with turnips.

Parish of *Tickhill*,

4958½ Acres of land from survey
2479¼ —— grass
2479¼ —— arable.

The letter-writer ſays, that one-third of the arable is turnip and fallow, one-third barley and oats, and one-third wheat and clover.

Pariſh of *Tinſley*,

260	Inhabitants
1435	Acres of land
570	—— graſs
15	—— waſte
300	—— woods
430	—— corn and clover
140	—— fallow.

Pariſh of *Wakefield*,

8192	Acres of land
6270	—— arable and graſs
1922	—— waſte.

The letter-writer ſays, the waſte is now incloſing.

Pariſh of *Warmfield*,

This pariſh contains two townſhips, viz.

1. Warmfield,	666	Inhabitants
	1517	Acres by actual ſurvey
	700	—— graſs
	618	—— arable
	177½	—— waſte
	11½	—— woods
	170	—— wheat
	60	—— barley
	70	—— oats
	58	—— beans
	90	—— clover
	180	—— fallow.

2. Sharleston, 176 Inhabitants
910 Acres estimated
360¼ —— grass
500¾ —— arable
50 —— waste
246 —— wheat
24½ —— barley
16¼ —— oats
31½ —— beans
59 —— clover
122½ —— fallow.

The letter-writer says, that the produce of this parish, on an average of the last eight years, from an exact account kept of the tithes, appears to be, per statute acre, as follows:

wheat per acre,	Winchester measure		18 Bushels
barley	-	-	32 do.
oats	-	-	36 do.
beans	-	-	18 do.

He adds, that most of the parishes of Agbrigg Wapentake may be estimated, if the quantity of corn raised be wanted.

Parish of *Weston*,

This parish contains two townships, viz.

1. Weston, 64 Inhabitants
1350 Acres estimated
655 —— pasture
360 —— waste
95 —— fallow
92 —— wheat
87 —— barley
56 —— oats
25 —— beans.

2. Askwith,

172	Inhabitants
1558	Acres eftimated
705	—— pafture
500	—— wafte
66	—— fallow
110	—— oats
117	—— wheat.
30	—— barley
20	—— beans.

The letter-writer fays, that this ftatement may be very erroneous, as the tenants were particularly referved in giving their communications, but that he has it not in his power to give a better.

Parifh of *Whifton*,

612	Inhabitants
2448	Acres of land
749	—— grafs
200	—— wafte
340	—— fallow
600	—— wheat and barley
220	—— clover
339	—— oats and beans.

Parifh of *Whitkirk*,

400	houfes containing from 1500 to 1600 fouls
3880	Acres of land
1100	—— meadow
1380	—— pafture
1180	—— arable
130	—— wafte
90	—— woods

820 Acres corn
60 —— turnips
300 —— fallow.

The wheat is generally more than half of the whole corn, the oats exceed the beans, and the beans the barley.

Parish of *[illegible]* in *[illegible]*,

1400 Inhabitants
69[illegible]0 Acres of land
1600 —— waste

The [illegible] in this parish is in general as 6 to 5, but in the [illegible] of R[illegible] Hay, the letter-writer says the [illegible] part of the whole. He says also, that [illegible] to parliament, for inclosing, was intended to have been made the following session.

Parish of *[illegible]*,

731 Inhabitants
5100 Acres of land
2563 —— grass
2537 —— arable
290 —— waste
818 —— wheat
383 —— oats
250 —— barley
2[illegible]6 —— beans
660 —— fallow.

This account taken partly from survey, and partly from estimate.

Parish of *Ilston* by *Belland*,

780 Inhabitants
3950 Acres of land estimated

3500 Acres graſs
450 —— arable
400 —— oats
25 —— wheat
25 —— fallow or ſmall pieces of beans or barley.

Pariſh of *Braithwell*,

500 Inhabitants
2750 Acres of land by ſurvey
1100 —— graſs by eſtimation
1650 —— arable Ditto.

The letter-writer ſays, the cultivation is ſo fluctuating that it cannot be preciſely aſcertained.

Pariſh of *Brotherton*,

900 Inhabitants
2110 Acres of land by ſurvey
1300 —— graſs
760 —— arable
50 —— woods
190 —— fallow or turnip
190 —— barley
190 —— clover
190 —— wheat.

The letter-writer ſays, that the quantity of open fields is not aſcertained; that the courſe of crops is generally as above ſtated; but that this ſyſtem of management is not univerſally adhered to. Some few acres of oats, beans, rapes, flax and woad, are occaſionally grown, but the quantity applied to the growth of any of theſe articles is ſo very ſmall, that he cannot exactly fix it.

Pariſh of *Burton Leonard*,

From 260 to 270 Inhabitants
1400 Acres of land
600 —— graſs
800 —— arable
260 —— wheat
290 —— barley, oats, and beans
250 —— fallow, turnip, and potatoes.

Pariſh of *Burghwallis*,

176 Inhabitants
1565 Acres of land as per regular ſurvey·
562 —— graſs
800 —— arable
203 —— waſte, of which 28 is highways
600 —— grain
200 —— fallow.

Pariſh of *Broughton*,

This pariſh contains two townſhips, viz.

1. Broughton, 160 Inhabitants
1580 Acres of land
50 —— waſte
15 —— in corn.

2. Elſlack, 125 Inhabitants
1150 Acres of land
400 —— waſte
44 —— in corn.

Pariſh of *Calverley*,

9900 Inhabitants.

Pariſh of *Otley*,

This pariſh contains 13 townſhips, viz.

1. Otley,

2360	Inhabitants
2291	Acres of land
2045	—— graſs
240	—— arable
34	—— wheat
122	—— oats
13	—— barley
9	—— beans
68	—— fallow.

2. Newhall with Cliſton,

194	Inhabitants
1380	Acres of land
1053	—— graſs
327	—— arable
50	—— wheat
200	—— oats
12	—— barley
15	—— beans
50	—— fallow.

3. Farnley,

231	Inhabitants
1303	Acres of land
721	—— graſs
477	—— arable
105	—— waſte
130	—— wheat
117	—— oats
57	—— barley
33	—— beans
140	—— fallow.

4. Lindley,

157	Inhabitants
968	Acres of land

	280	Acres of grafs
	288	—— arable
	400	—— wafte
	49	—— wheat
	122	—— oats
	27	—— barley
	18	—— beans
	72	—— fallow.
5. Little Timble,	50	Inhabitants
	458	Acres of land
	268	—— of grafs
	90	—— arable
	100	—— wafte
	6	—— wheat
	48	—— oats
	6	—— barley
	2	—— beans
	28	—— fallow.
6. Derton,	180	Inhabitants
	2581	Acres of land
	1398	—— grafs
	183	—— arable
	1000	—— wafte
	60	—— wheat
	40	—— oats
	8	—— barley
	15	—— beans
	60	—— fallow.
7. Burley,	705	Inhabitants
	3662	—— of land
	1188	—— grafs
	476	—— arable

	2000 Acres waſte
	84 —— wheat
	219 —— oats
	15 —— barley
	29 —— beans
	127 —— fallow.
8. Menſtone	150 Inhabitants
	750 Acres of land
	370 —— graſs
	330 —— arable
	50 —— waſte
	35 —— wheat
	165 —— oats
	40 —— barley
	90 —— fallow.
9. Hakeſworth,	220 Inhabitants
	1873 Acres of land
	651 —— graſs
	422 —— arable
	800 —— waſte
	60 —— wheat
	190 —— oats
	51 —— barley
	11 —— beans
	110 —— fallow.
10. Eſholt,	224 Inhabitants
	417 Acres of land
	235 —— graſs
	182 —— arable
	40 —— wheat
	35 —— oats
	30 —— barley

P 2

	16	Acres beans
	61	—— ſallow.
11. Bramhope,	231	Inhabitants
	1050	Acres of land
	250	—— graſs
	380	—— arable
	420	—— waſte
	30	—— wheat
	160	—— oats
	80	—— barley
	10	—— beans
	100	—— ſallow.
12. Pool,	202	Inhabitants
	774	Acres of land
	429	—— graſs
	345	—— arable
	73	—— wheat
	160	—— oats
	71	—— barley
	16	—— beans
	79	—— ſallow.
13. Baildon,	2220	Inhabitants
	2234	Acres of land
	777	—— graſs
	543	—— arable
	913	—— waſte
	98	—— wheat
	257	—— oats
	46	—— barley
	32	—— beans
	110	—— ſallow.

Parish of *Pontefract*,

6625 Inhabitants by estimation
5112¼ Acres of land
2160½ —— grass
325 —— common pasture
25¾ —— waste
484¼ —— fallow
458 —— wheat
265¼ —— maslin
638¼ —— barley
277½ —— oats
146 —— beans
121¼ —— turnips
53¼ —— potatoes
16 —— rape
2½ —— cabbages
138¼ —— nurseries, liquorice, gardens, and orchards.

This letter-writer says, there are also let upon lease from the crown and included in Pontefract Park Ville, 1019¼ acres, of which is in grass 200 acres, in fallow 170 acres, and in corn 649¼, besides 47 acres called King's close, of which 23 acres are in grass, 5 acres in fallow, and 19 acres in corn.

Parish of *Royston*,

1690 Inhabitants
8178¾ Acres of land
2948¼ —— grass
4993½ —— arable
237 —— waste
1292¼ —— fallow
3701¼ —— in corn.

Parish of *Rothwell*,

8727 Acres of land
3817 —— grass
4680 —— arable
230½ —— common
383½ —— clover.

Parish of *Sandal Magna*,

2300 Inhabitants
6480 Acres of land by estimation
2168 —— grass
3122 —— arable
600 —— waste
590 —— woods
926 —— wheat
469 —— oats
367 —— barley
343 —— clover
180 —— turnips
152 —— beans and pease
12 —— potatoes
673 —— fallow.

Parish of *Sheffield*,

This parish contains 6 townships.

1. Sheffield,

7351 houses at 4½—33079 Inhabitants
3436½ Acres of land.

2. Ecclesall Bierlow,

1071 houses at 4½—4819 Inhabitants
4180 Acres of land.

3. Beightside Bierlow,

822 houses at 4½—3699 Inhabitants
2680 Acres of land.

4. Attercliffe and Darnal,

500 houses at 5—2500 Inhabitants

1119½ Acres of land exclusive of houses, yards, and Attercliffe green

217 Acres waste.

5. Upper Hallam,

105 houses at 4½—472 Inhabitants

5086 Acres of land

3150 —— waste.

6. Nether Hallam,

188 houses at 4½—846 Inhabitants

1877 Acres of land

25 —— waste.

The letter-writer states, that the above is part from survey, and part from estimation; that the empty houses are included, excepting those in Attercliffe and Darnal. He adds that of the 3436½ acres in No 1, 700 are occupied by the town of Sheffield; that 3450½ acres in No 2, are old inclosures, and 730 acres new inclosures; that the waste land in No 4 consists of Attercliffe common 177 acres, and Darnal common 40 acres. And that the 25 acres entered as waste in No 6, is inclosed or about to be inclosed. He farther says, that most part of the parish of Sheffield, especially the lands near the town, is chiefly in grass, but is not suffered to remain many years without being plowed, and two or three crops of corn being taken from it, generally wheat and oats.

No XI.

ACCOUNT of the Parish of DRAX, transmitted by JOCELYN PRICE, Esq; an Active and Intelligent Magistrate.

The Parish of Drax, 1796.		Townships. Camblesforth.	Long Drax.	Newland.	Drax.	The Parish.
Number of Acres in Arable Land	Wheat	112	210	278	156½	756½
	Oats	87	190	295	196	768
	Barley	18	30	10	15¼	73¼
	Rye	23				23
	Maslin	34	6	14	43	97
	Beans	9	86	180	69½	344½
	Turnips	46	10	2	9¼	67¼
	Potatoes	33	48	54	32	167
	Flax	27	2	15	30	74
	Rape		10		43	53
	Teasels	7				7
	Fallow	130	198	290	130	748
	Clover	29	31	35	16	111
Total Number of Acres	Arable	555	821	1173	741	3290
	Meadow	71	134	130	115	450
	Pasture	260	342	400	240	1242
	Woodland	70	25	8		103
	Waste or Common	695				695
	Gardens	2	1½	1	2	6½
	Orchards	2	4	20	3	29
Total number in each Township, and the Parish		1655	1327½	1732	1101	5815½
	Population	179	167	190	213	749

Produce	Cannot be afcertained.
Cuftomary hufbandry	In general two crops and a fallow.
Peculiar hufbandry	Some farmers ufe the drill.
Manures	Lime, fold & Hull manure.
Acres capable of drainage	All drained except Camblesforth common.
Mode of drainage	Wide drains, and cloughs in the river's bank.
Number of Acres capable of being protected by embankments	All protected, except in very great frefhes or floods.

OBSERVATIONS.

An enclofure of Camblesforth common would be a great improvement; and if all the lands adjoining the rivers Oufe and Aire were warped, it would confiderably improve them, and enhance their value.

www.ingramcontent.com/pod-product-compliance
Lightning Source LLC
Chambersburg PA
CBHW021345210326
41599CB00011B/752

* 9 7 8 3 7 4 1 1 5 3 2 5 9 *